John A. Whitehead

Energy Flow and Earth

How Earth Works

 Springer

John A. Whitehead
Department of Physical Oceanography
Woods Hole Oceanographic Institution
Woods Hole, MA, USA

ISSN 2191-589X   ISSN 2191-5903 (electronic)
SpringerBriefs in Earth System Sciences
ISBN 978-3-031-62693-7   ISBN 978-3-031-62694-4 (eBook)
https://doi.org/10.1007/978-3-031-62694-4

SpringerBriefs in Earth System Sciences

Series Editors

Gerrit Lohmann, Universität Bremen, Bremen, Germany

Justus Notholt, Institute of Environmental Physics, University of Bremen, Bremen, Germany

Jorge Rabassa, Labaratorio de Geomorfología y Cuaternar, CADIC-CONICET, Ushuaia, Tierra del Fuego, Argentina

Vikram Unnithan, Department of Earth and Space Sciences, Jacobs University Bremen, Bremen, Germany

SpringerBriefs in Earth System Sciences present concise summaries of cutting-edge research and practical applications. The series focuses on interdisciplinary research linking the lithosphere, atmosphere, biosphere, cryosphere, and hydrosphere building the system earth. It publishes peer-reviewed monographs under the editorial supervision of an international advisory board with the aim to publish 8 to 12 weeks after acceptance. Featuring compact volumes of 50 to 125 pages (approx. 20,000—70,000 words), the series covers a range of content from professional to academic such as:

- A timely reports of state-of-the art analytical techniques
- bridges between new research results
- snapshots of hot and/or emerging topics
- literature reviews
- in-depth case studies

Briefs are published as part of Springer's eBook collection, with millions of users worldwide. In addition, Briefs are available for individual print and electronic purchase. Briefs are characterized by fast, global electronic dissemination, standard publishing contracts, easy-to-use manuscript preparation and formatting guidelines, and expedited production schedules.

Both solicited and unsolicited manuscripts are considered for publication in this series.

Dedicated to the Geophysical Fluid Dynamics Summer School

Preface

This relates to my personal experience as I watched mankind's discovery of the basic understanding of how the flow of energy acts as a giant architect for many large things on Earth. These "things" include the giant plates, volcanos, continents, ocean basins, Earth's magnetic field, our atmosphere, and ocean circulation. My journey as a scientist taught me how energy flow produces overall support while mechanics and chemistry provide invisible skeletons that hold them together. Energy flow is invisible, and it takes many forms. Two examples are an electric current that brings energy into the house and sunlight that provides a flow of heat to warm us when we are cold. My specialty led to studies of the effect of energy flow on our planet that involved work both in a laboratory and in the Earth. I am excited to write about this, because much of this invisible skeleton was uncovered in my lifetime.

My trip started when I became a graduate student in Engineering and Applied Science at Yale in 1963 after being trained as a mechanical engineer at Tufts University. After completing my thesis in fluid mechanics and receiving a PhD in 1968, I went to UCLA to learn about geophysics and planetary science and then to the Woods Hole Oceanographic Institution to learn about the ocean and marine geology, where I have been ever since. I have been extremely fortunate to have experienced personal contact and to have known a large percentage of the people involved in the discoveries in this book. I worked with at least two hundred of them! I attended countless seminars and read hundreds of research papers. Many of the people who developed these ideas have now died and quite frankly, I want to write this view down before all of us are gone. This book expresses my sense of obligation to all of them.

This book is my own view from my own experience rather than a fastidious history. I don't include large lists of names, biographical snippets, and citations of all the contributors. Numerous talented contributors are generally cited in many books, and there are countless review papers about topics discussed in this book. As a first start at describing my experience, I have found that many people are left out of any list of the famous. Some examples are technicians left out of the list of authors; volunteers; people who simply were in a laboratory or classroom at a moment when they made a remark that made all the difference; students who were simply obscured by efforts by more senior people who swiftly followed them; students who did a great job but then went into another field so that their names are not commonly recognized anymore; scientists who made the first discovery but were not well-known because of missing citations in later papers; people who had political or financial constraints that made their results obscure; and people who unfortunately had a second person following their work who was more eloquent and quick to claim the credit. I can name examples of each of these. Each chapter has citations about significant contributions and key books. Although they are a good guide for the historian, I'd like to paraphrase ("with tongue in cheek") a wonderful friend, Ed Spiegel who said something like "I kept leaping up because the giants' shoulders were blocking the view". We all did this together!

In this book, I want to tell a central part of this story that describes giant flows of rock, water, air, ice liquid metal, and magma. We learned that the global flow patterns of these flows are constructed or at least strongly influenced by natural processes with an invisible underpinning of energy flow. We can visualize the flows as being components of large engines, and the story includes the flow of energy propelling them. Therefore, this book includes the principal concepts of energy flow. The invisible flow of energy not only runs these giant engines, but it also takes an active part in producing and maintaining the flows and structure of the engines. Many other things around us are produced and driven by a flow of energy, too. For example, all of nature is driven by the flow of energy. Every living thing needs energy flow to produce and reproduce itself and keep itself going. This is true of bumblebees, birds, sunspots, tornados, mountains, oceans, rainstorms, and hurricanes. It is also true in many other aspects of our existence such as city traffic, politics, economics, and kings. They concern ideas I learned as a scientist. Some of them were discovered as I watched.

We describe natural flows and think about how they were started and about the power that propels them. Like the gasoline driving a running engine, heat flow provides the power to drive many flows on Earth and provides a ghostlike figure behind a skeleton of mechanics and chemistry holding the flows together. Both some material substances and the flow of energy are needed to explain Earth's giant engines. Usually, the latter has been ignored. My own ghost is my food and metabolism, and my skeleton includes all the structures and functions of my own body. When my ghostlike internal energy flow stops, I will die and decay away. But the ideas in this book won't. Hence, time to write!!

Anyway, here is how it all works.

Woods Hole, USA   John A. Whitehead

Contents

1   **An Introduction to Energy Flow and the Discovery
    of the Great Plates** 1
    1.1   Energy Flow 1
    1.2   What Was Known About the Deep Earth 5
    1.3   New Data Introduced the Moving Earth 7
    References 14

2   **Spontaneous Production of Flow Patterns—Cellular
    Convection** 15
    2.1   A Simple Demonstration of Convection Cells 15
    2.2   Rayleigh's Number and Linear Stability 19
    2.3   Dimensionless Numbers and Scaling 20
    2.4   Instability—The Growth of Practically Nothing 21
    2.5   Finite Amplitude Theory 28
    2.6   Large Ra Results and Boundary Layer Flows 34
    2.7   Summary 39
    References 40

3   **Energetics of Mantle Convection** 41
    3.1   Heat Flow and Mantle Convection 41
    3.2   Cellular Convection and Plate Motions 47
    3.3   Energy Flow 52
    References 57

4   **Energy Flow and Magma Generation** 59
    4.1   Regions of Magma Production on Earth 59
    4.2   Flow Localization in the Laboratory 64
    4.3   Arrivals and Eruptions, Flood Basalts, and Extinction Events 72
    4.4   Energy-Flow Rates of Mantle Convection 75
    4.5   The Calculations 79
    References 80

5   Power and the Building of Continents, Mountains,
      and the Ocean Basins 83
      5.1   The Distribution and Structure of Continents 83
      5.2   The Mechanics of Continents and Ocean Basins 86
      5.3   The Equilibrium Areas and Thicknesses of Continental
             Crust and Water 89
      5.4   Interaction of Floating Plates and Layers with Convection
             Cells 92
      5.5   Summary of Energy-Flow Rates for Continents and Ocean
             Basins 97
      References 98

6   Energy Flow and Producing the Earth's Magnetic Field—The
      Dynamo 101
      6.1   Mechanical Dynamos 101
      6.2   Experiments and Theory 104
      References 109

7   Power That Drives Circulation of the Atmosphere and Oceans 111
      7.1   The Atmosphere 111
      7.2   The Oceans and Their Surface Circulation 119
      7.3   Deep Ocean Energy Flow 120
      References 129

8   Speculations About Energy-Flow Structures in Nature
      and in Human Activities 131
      8.1   The Use of Energy Flow by Social Groups 131
      Reference 133

9   A Tutorial Review, Simple Energy-Flow Calculations,
      and Basic Definitions 135
      9.1   Energy and Work 135
      9.2   Energy Flow and Its Rate 138
      9.3   Quantification of Properties, Units 142
      9.4   Large Numbers 142
      9.5   Dimensionless Numbers 143
      Reference 146

10   Summary 147

About the Author

John A. Whitehead studied mechanical engineering and intended to enter human engineering and make machines easy for people to use. Instead, he became enchanted by the challenges of fluid dynamics and went to graduate school. A student project on the fluid mechanics of thermal convection led to a life's work in studying natural fluid flows, starting at the Institute of Geophysics and Planetary Physics at UCLA, followed by about 50 years in the Geophysical Fluid Dynamics Laboratory at Woods Hole Oceanographic Institution. His experimental work on convection helped us to understand the nonlinear character of stability theory for thermal convection. He has also done fundamental experimental work on the dynamics of flows in volcanic conduits in the Earth's mantle, in deep, hydraulically controlled oceanic flows in straits connecting ocean basics, the fluid mechanics of coastal currents, and on multiple equilibria in fluid dynamics concerning oceanic thermohaline circulation. Work was also spiced up with theory, field trips, ocean cruises and, of course, countless seminars including the Institution's Geophysical Fluid Dynamics Summer Study program. He is a Guggenheim Fellow, a Fellow of the American Physical Society, the American Meteorological Society (Henry M. Stommel Research Award, 2007), and the American Geophysical Union (Ewing Medal, 2014) and a member of the American Academy of Arts and Sciences.

Chapter 1
An Introduction to Energy Flow and the Discovery of the Great Plates

Abstract This chapter begins with a discussion about the flow of energy. The story begins with a short history of the discovery of mantle convection. In the early part of the twentieth century, the ideas that continents slowly move around on the surface of the Earth was put forth, but after a couple of decades the idea remained stagnant. Then it was found that these continents were on great rigid moving plates that are formed at ocean ridges. The plate material in the seafloor moves away from the ridges and ultimately plunges back into the Earth at ocean trenches. This chapter tells of how the primary features of the great plates were discovered and how sinking slabs of cold ocean floor within the mantle of the Earth were detected. These were shown to drive the plates in the 1960s and early 1970s.

1.1 Energy Flow

In this book, we describe the energy flow that drives numerous things in our natural world. Energy flow occurs everywhere. We put gasoline in an automobile and the motor takes us on a trip by burning the fuel. The gasoline flows into the engine to propel us. We burn fossil fuels to produce heat. We use fuel to propel machines that provide food, heating, transportation, and almost everything else that runs our modern civilization. Unlike many books and essays that are coming out now, this book is not about humanity's energy crisis. It is instead a primer about how the flow of energy, generally in the form of heat, naturally moves and forms things on Earth. Unlike an automobile, these "giant flows" are not made by people but are even constructed or at least strongly modified by the flow of the energy. They are truly natural engines. Some examples of these giant engines are the continents, ocean basins, storms, tornados, volcanos, mountains, rivers, ocean waves, and even our magnetic field.

Energy flows everywhere, and we experience aspects of the flow all the time. We feel warmth from a heater in cold weather, and a cool breeze removes heat from our face on a hot day. Beyond our personal reach, we marvel at the giant energy expended by a large steamship, the awesome power of a train locomotive, and the immense

© The Author(s), under exclusive license to Springer Nature Switzerland AG 2024
J. A. Whitehead, *Energy Flow and Earth*, SpringerBriefs in Earth System Sciences,
https://doi.org/10.1007/978-3-031-62694-4_1

cascade and turbulence of a waterfall. Very high winds show us great forces from the invisible drag of air. We marvel at a giant crashing wave or a bolt of lightning. Projectiles from the wind can scare us. We hop into a car, bus, train, or airplane to travel at speeds that were not even imagined by Benjamin Franklin. Electric motors operate everywhere driven by an invisible flow of electrons. Energy flows invisibly all around us, and the flow of energy is vital to all we do.

Despite its continual presence, the flow of energy is not generally appreciated, probably because it is not easy to visualize and understand. We only know about energy and its flow as a concept or a feeling. Energy flow is invisible even though it is present! Nobody holds flowing energy in their hand or pokes it or paints a picture of it or photographs it. For example, with electricity we don't see all those electrons flowing through wires, but we see the video screen or hear the radio. We don't see the heat flowing into our frying pan, but we certainly know if it gets hot.

Energy is contained in all matter, and it can be stored and transferred in many forms. When energy flows into a region, it can change the energy within the region in various ways. It can make the temperature increase, it can produce a compositional change such as solidification, by melting it, it can start a chemical reaction, or it can simply increase kinetic energy. If gravity (or another body force) is present, energy flow can change potential energy. Energy flow can involve an electric current or create a change in a magnetic field. Energy often changes its form in complex ways.

A simple example of changing forms of energy starts with cold firewood inside a fireplace. After igniting the wood, the chemical energy stored in the carbon of the wood links up with the oxygen in air to release heat and form carbon dioxide. The air leaving through the chimney is much warmer than the ambient air, so the fire exports this heat along with the carbon dioxide. The fire also exports heat by radiation. Therefore, the chemical energy of wood and air has changed to heat. The remaining materials, mostly carbon dioxide and ash, have lower chemical energy. In this book, we will look at power budgets that are in a fixed region just like this fireplace, but our regions will be laboratory fluids or portions of Earth, such as our atmosphere, ocean, mantle, and core. Some of these concepts are being applied to other planets, too.

This book concerns energy-flow structures. The fireplace itself is not strictly an energy-flow structure, because people constructed the fireplace, prepared the wood, and ignited it. Most mechanical machines that we deal with, and many of our activities (such as cooking) are made by people. In contrast, I think of the fire itself and the circulation of heat associated with it as an energy-flow structure. The energy flow within the flame converts chemical energy to three forms. The first is kinetic energy production of the air flow. The second is heat flow away from the flame by both thermal conduction and convection. The third is radiation of the heat energy. The flame only exists after being started. For a flame, an elevation of temperature above a certain value is needed to start it. After ignition, the flame produces the conditions to keep itself going. Chemists can give us a formula that expresses the required temperature for a fire to be self-sustaining. This formula is a criterion for the existence of this energy-flow structure. Some energy-flow structures start spontaneously after the ambient conditions slowly change to be moved into a situation that fits the criterion. If we made the bricks or rocks of the fireplace hot enough, the fire would start without a match.

Energy flow is more than a servant. Did you know that not only do these giant flows that we see on Earth come from a flow of energy but also that most of them will retain their shape only with continual energy flow? One might call the structures associated with these giant flows "Earth's engines", even though no factory made them. Energy flow is involved in the construction of many of the gigantic things that we consider part of Earth: our land; the sea; our air; the wind; ocean waves; ocean currents; and even our mysterious magnetic field. All of nature is driven by the flow of energy, too, although smaller engines in biology are not included in this book. Every living thing needs energy flow to make itself grow from a tiny seed or egg and to keep itself alive. When energy flow stops in every living thing, it dies and decays away.

So, precisely what is energy flow? It is connected to a concept in mechanics that is literally called "work". This concept has a precise scientific definition. A tutorial example is given in Chap. 9. The mechanical definition is unlike our common usage of the word. If I say to a small child, "How Earth works!" he or she might imagine a cartoon with an animated planet and a cartoon of Earth that is hard at work doing something. To someone in the working community, the concept of labor comes to mind. On the other hand, if I say this to an engineer, physicist, or person interested in how nature works, that individual might imagine that I will explain exactly how something mechanically works. Likewise, a chemist might think I am going to explain how work and energy are involved in chemical reactions that produce all the materials we use (including for example gasoline). A geochemist might think I write about Earth's past and how the oceans and land as we know them were formed (including, for example, rocks, water, and air). Energy flow is readily measured, and, in this book, it will be expressed in Watts (with the abbreviation W). More aspects of the units are discussed in a tutorial in Chap. 9.

This book involves all these roles involving work and energy, but we go further. I will show how energy flow and the rate of work (using the scientific definition of work) PRODUCES and SHAPES things. In fact, it helps to produce and shape almost everything in our natural world. Energy flow can act like the skeleton sketched in Fig. 1.1a lurking behind the heat ghost.

Energetics associated with classical physics is so well studied that many physicists find it rather dull. However, in this book we will go one step beyond this simple energy flow and dissipation. We don't impose a prescribed machine. Instead, a specific mechanical machine is formed in the process.

What???

Yes, we show how over the past 150 years it slowly was discovered that by using only simple classical physics, we can understand how many new things are formed as energy flows. The "new things" are not atoms, molecules, or any sort of particles alone, nor are they locomotives or steamboats or airplanes or cars or roller coasters. Instead, they are energy-flow structures that form only when there is a flow of energy in and out of a region, like the flame we already mentioned. It's like an automobile being manufactured that includes the inventor, the factory workers, and the mechanic.

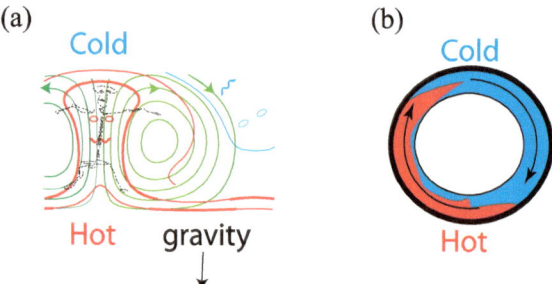

Fig. 1.1 **a** The heat ghost and an underlying skeleton. A hot thermal (red isotherms) rises from a hot bottom next to a cold sinking downdraft from a cold top. The ghost represents energy flow, and the skeleton-like framework is some sort of mechanical law governing a response (green streamlines). **b** Buoyancy-driven heat flow in a tube of water driven by heating from below

There are various technical terms to refer to energy-flow structures depending on the community interested in them. Some examples of other terms are dissipative structures (chemical physics); finite-amplitude flow (applied math); nonequilibrium structures (physics); or self-organized structures (physics and the media).

There is a hint about how energy-flow structures operate the overturning flow in the tube of water heated from below that is sketched in Fig. 1.1b. It reveals why I am continually hypnotized at how physics and mathematics build upon each other. Mathematically, there are four aspects of even the very simple flow in Fig. 1.1b. First, there is the possibility of either zero flow at all, just hot water along the bottom and cold at the top. Second, the fluid mechanics and heat transfer principles dictate that here is a minimum criterion for the existence of the flow to grow from zero. The criterion depends on several factors such as the temperature, the shape and size of the passageways, the expansion of the liquid with temperature, the strength of gravity, the thermal conductivity of tube walls liquid, and the viscosity of the fluid. We will show how these are grouped together to make a formula to express a criterion that must be exceeded for flow to exist. Third, there might be a need for a large push to switch from zero flow to the full flow shown there. If the flow were intelligent (which it is not), we could anticipate that the fluid has flow options, but here I want simply to think of these options as mathematical branch points between no flow and flow. The location of the branch points depends on the value of the criterion we mention above, or even on other criteria. Fourth, the rate of heat transfer (and therefore the rate of release of potential energy) for the situation with no flow has a different value than the one with flow present. I like this example because it shows the relation between physics and math. Numbers 2 and 4 are physical criteria, and 1 and 3 are mathematical aspects of branch points.

It is easy to imagine more complex examples than Fig. 1.1b that have even more branches. Warning: Numerous branch points can produce confusion! Branch points are responsible for the complexity and uncertainty about many commonplace aspects of our world: weather is variable, turbulence exists, geology is complex, nature is full of surprises, and the future is hard to predict.

In the rest of this chapter, we describe the discovery of an immense flow pattern produced by and even formed by energy flow in the mantle of Earth. In the following chapter, we review our understanding of how that giant pattern formed. Then, in Chap. 3 we give the story of the speeds that are involved. In the following chapters, we tell of the existence of some other energy-flow structures on Earth. Here we go!

1.2 What Was Known About the Deep Earth

When I was a graduate student at Yale in the mid-1960s, I attended a seminar by Professor E. R. Oxburgh from Oxford University who showed on the blackboard some calculations that linked finite-amplitude convection cells (see Chap. 2) to continental drift on Earth. The work was done jointly with Professor Donald Turcotte of Cornell University (Turcotte and Oxburgh 1967). The structure and flow were extremely suggestive of how flow inside the Earth occurred. The work had a major impact on me but also on the entire field! My curiosity clearly turned toward the Earth.

I purchased some books about what is known about Earth. Much information about material below our surface is found in seismology, which is the science of the propagation of waves caused by earthquakes. These waves propagate in the upper parts of Earth. They were even known in ancient times. Early instruments in ancient China consisted of balanced balls coupled to pendulums that could indicate the direction of a seismic wave and thus indicate the direction of origin of an earthquake. Instruments to measure seismic waves gradually become more precise over time, so that by 1900 the data from the instruments, that by then were named seismometers, indicated that the Earth had internal reflection layers. There are two kinds of waves. One kind is compression waves, like sound. The other kind is elastic waves, like the waves in gelatin and there are both surface and internal elastic waves. As instruments became better and more numerous, their effective resolution extended to deeper and deeper depths, so that the layers became more clearly defined at many depths. By 1960, the knowledge of the deep Earth from seismometers allowed sketches in textbooks like Fig. 1.2a.

The topmost of these reflection layers is at depths between 50 and 100 km deep. This was located at the base of the upper lithosphere, a stiff cold region. At the top of the lithosphere is a layer of crust that is too thin to be seen here. The crust exhibits a lot of lateral variation and was the concern of numerous seismologists, but it is localized and is the focus of a later chapter. Below the lithosphere is a region called the mantle that has regions of relatively uniform seismic propagation speeds. The mantle has two layers. They are separated by an internal reflection layer at roughly 660 km below the surface that is generally thought to be a transition to a different crystal structure below that depth. The fact that both elastic and pressure waves move through the mantle led to the conclusion that the mantle is mostly a solid, elastic material. The base of the mantle is at about 3486 km radius from Earth's center, and it lies above a liquid "core" that supports sound waves but not elastic waves. Various estimates

(a)   (b)

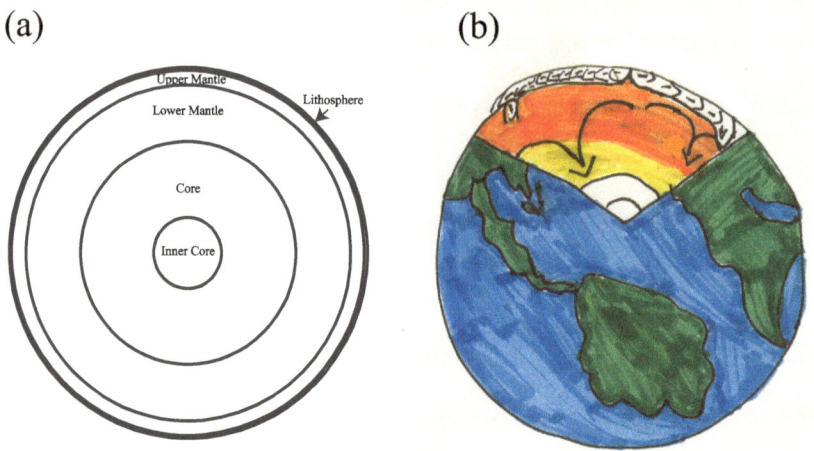

Fig. 1.2 **a** Seismic reflection layers within the Earth known in 1960. **b** The new view of Earth's interior. Artwork by Charlotte Nangle

of the mass of the Earth led to the belief that the core is mostly liquid iron. By the 1930s, seismological measurements and analysis of the elastic and pressure waves had become so precise that a second reflecting surface from a solid "inner core" was found at a radius of about 1200–1250 km. The inner core surface is not smooth, and the traversing of the bumps under a spot on the Earth's surface led to different arrival times so that the differential rotation of the inner core was discovered.

The layered mantle in Fig. 1.2a was the fundamental picture of the interior of the Earth in the early 1960s, although lateral differences were known near the surface. Of course, the continents and ocean basins were fundamentally different, and high-resolution seismology, along with acoustics, gave evidence of differences between mountains, flat continent areas, and ocean floors to depths of up to 100–200 km.

Great moving spreading centers were being discovered on the ocean floors and this led to the different picture in Fig. 1.2b. There were new and clearer ideas about what is going in the mantle and core. Excitement was great by the time of the Oxburgh lecture, because by the mid-1960s the motion solved several puzzles that had long challenged Earth scientists. The greatest puzzle that was solved concerned the continental drift question. This has arisen because several lines of evidence were consistent with the Atlantic Ocean forming during the past 200 million years. An entire book, with multiple editions, had been composed in the early 1900s by the meteorologist Alfred Wegener listing many lines of evidence: Fossil records showed that the evolution of land animals matched each other on the two sides of the Atlantic Ocean before 200 million years ago and did not match later. In addition, the shapes of the continent edges, (the geological edge is called the continental shelf break) in North and South America matched almost perfectly with the shapes of the Europe and African edges. Moreover, smaller geological features such as rock types and mountain chains also matched. Numerous other aspects from magnetic dating of rocks and the matching of geologic structures on the two sides of the Atlantic Ocean are also described. Although the geologist Arthur Holmes suggested that convection cells existed within the mantle and moved the surface continents back and forth, no firm collection of data was available to clarify the concept. The puzzle of how an entire ocean basin could form, or how continents might possibly plow over the mantle was a mystery that prompted many Earth scientists to reject, or at least to ignore, the hypothesis of "Continental Drift".

1.3   New Data Introduced the Moving Earth

Unknown to me as an engineering graduate student, new data in the late 1950s had suggested that the Atlantic Ocean has been steadily spreading apart in the middle for millions of years. The data came from an instrument to measure magnetic variation at sea. The "magnetometer" was originally developed for submarine or mine detection and was towed behind either ships or aircraft to produce profiles of Earth's magnetic field as they moved. Small variations in strength from the large overall magnetic field were discovered and then mapped out by marine geologists in the 1950s. These variations are produced by the very small magnetism that is frozen within ocean rocks on the ocean floor. Early studies on the land of rock magnetism had revealed that Earth's field experiences reversals in polarity every tens of thousands of years, with the North pole of a magnet changing to the south and vice versa. Between 1958 and 1961, it was noticed that the ocean floor magnetic variations have similar patterns on both sides of ridges. The first ridge that was mapped was the Juan de Fuca Ridge offshore of Seattle and the next was the Mid-Atlantic Ridge (This ridge occupies the middle of the Atlantic all the way from Iceland in the north to the middle of the ocean near Antarctica in the south). From 1960 to 1964, as progressively more magnetometer traverses of the ridges were made, excitement quickly grew as the patterns were found to have the same shape on both sides of the ridge axis almost everywhere. Even more exciting was that the patterns extend all the way to the continents on both sides, and in some cases the patterns of the shapes were even the same in the Atlantic and Pacific oceans. Although the interpretation was originally controversial that these patterns are evidence of seafloor spreading away from mid-ocean ridges, the controversy died out. Over entire regions the patterns are often stripes. Figure 1.3 is a modern map of the magnetic field in one section of the North Atlantic. Iceland and Greenland are near the top and the mid-Atlantic Ridge is highlighted by a thick black dashed line. The light and dark patches of grayness indicate the direction of the magnetic field frozen in the rock. Reversals appear as white that rims the gray and black patches.

How were these magnetic patterns produced and what did that imply for the Earth? It was known that as lava rises and cools, there are trace materials that are ferromagnetic (the atoms are essentially tiny permanent magnets). All ferromagnetic material (such as iron) is magnetic below a fixed temperature called the "Curie temperature"

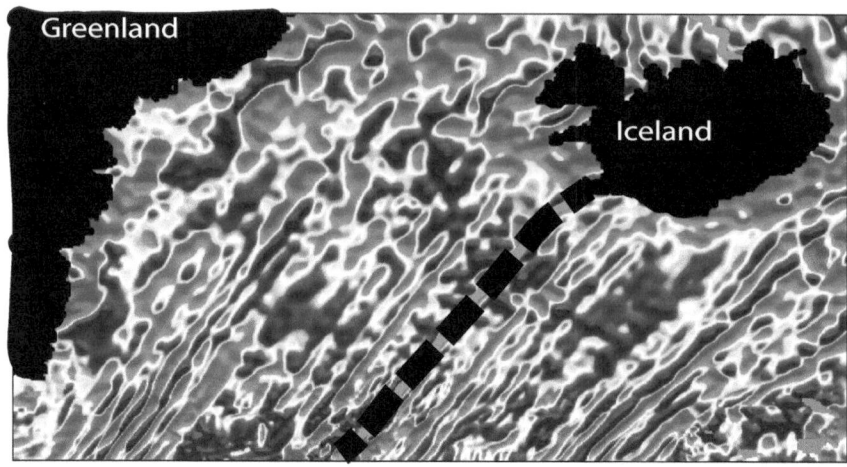

Fig. 1.3 A modern map of the magnetic field that is frozen into the seafloor of the north Atlantic that in many cases forms stripes. Greyscale indicates the magnetic anomaly. Greenland is in the upper left, Iceland in the upper right, and the mid-Atlantic ridge is the thick black dashed line. This image adapted from a sea level color-relief image from EMAG2_inage_V2.jpg available at https://www.ncei.noaa.gov/products/earth-magnetic-model-anomaly-grid-2

but it has no magnetism above the Curie temperature. When lava solidifies, these minerals form crystals that tend to align with the prevailing magnetic field as it cools below the Curie temperature. The oceanic crust and perhaps some of the mantle below the crust possess a small value of magnetism that indicates the direction and strength of the magnetic field when it solidifies. In addition to all this, other, land-based measurements of trace magnetism in rocks had shown that the magnetic field reverses direction roughly every 200,000 years. Therefore, the outrageous (at first) interpretation that comes out of the data is that the Earth's mantle is rising and splitting apart at the center of the North Atlantic Ocean so that the surfaces (that is, the solid ocean floors) are moving away from each other. The material that fills in the crack as they move away cools and produces some magna that solidifies. The ferromagnetic material records the magnetic field strength and direction when it solidifies. Figure 1.4 is a sketch of the concept.

Throughout the 1960s, measurements in different parts of the world found seafloor stripes and revealed some features that were common to all of them. The ocean floor seemed to be like a collection of gigantic tape recorders of the magnetic field (Everyone knew what a tape recorder was in those days!). As data from around the world were obtained, it was found that the detailed time series patterns worldwide of the reversals were similar, so the history of the magnetic field reversals became known in detail. Different spacing distance of lines with the same patterns was attributed to different spreading rates. Not everyone agreed with this interpretation, but when spreading rates were gathered from magnetic stripe data in locations all over the globe, a surprising conclusion was reached; that the ocean floor moves like a

Fig. 1.4 Sketch of how the reversal of a planet like Earth's field at three times is recorded as ferromagnetic material rises, solidifies, and spreads apart in spreading centers

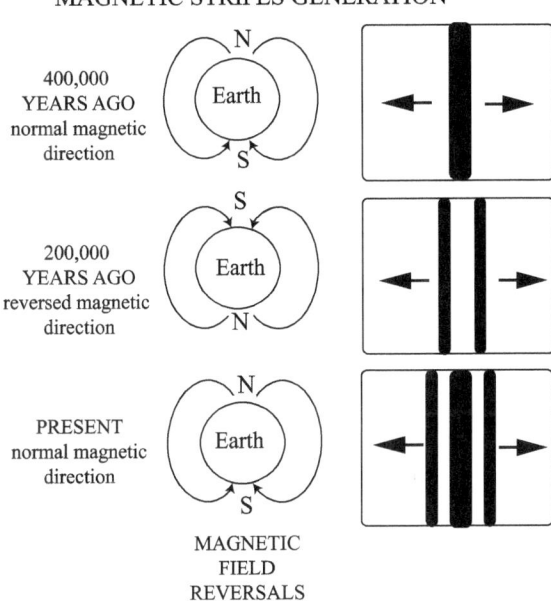

MAGNETIC STRIPES GENERATION

400,000 YEARS AGO normal magnetic direction

200,000 YEARS AGO reversed magnetic direction

PRESENT normal magnetic direction

MAGNETIC FIELD REVERSALS

collection of solid plates. The plates soon became well-known and many were given names. The relative speeds and direction of motion between plates were documented. The motion for all plates was (and still is) known only with respect to each other and the global average is not necessarily the average speed over the deep mantle.

In summary, to a first approximation, each large slab of ocean floor moves as a rigid plate. The technical word for this topic is "plate tectonics" (plate structures), although some early studies called them blocks. In addition to the magnetic data, other data and calculations reinforced the concept. For example, the motions of the plates were reconstructed using relative poles of rotation between pairs of plates projected on a sphere (The motion of any solid that is both moving and rotating can be shown to be equivalent to pure rotation about a point in space called the "Euler pole"). For plates on Earth, the Euler poles were determined between pairs of plates, although it is important to realize that the actual speed of the entire mantle with respect to the plates was not known. Global reconstructions of such poles made it increasingly clear that the plates were almost perfectly rigid with most deformation occurring only along the margins of the plates. Other evidence of the solid plate concept was found, too. For example, large linear fractures tend to align with the shear direction between neighboring plates everywhere on the spherical globe. Despite the remarkable agreement, not everyone agreed with the concept of rigid plates at first. When I was job-hunting in about 1970, the world of Earth scientists had many skeptics!

Marine geologists continued to accumulate new evidence that the ocean floors are created at ridges. In fact, the evidence has proven to be so convincing that ridges are now widely described as spreading centers. Ocean sediment data helped us to understand the story. The erosion from land, along with biological activity in the oceans, has the result that the ocean floor is typically covered by sediments. Their thickness ranges from hundreds of meters to a few kilometers, depending on location. In the early 1970s, dating of the ocean sediments was completed along continuous lines from ridge to old basin, first in the Atlantic and later in the Pacific. Two completely independent dating methods were used. 1. Identifying certain types of shells and 2. Measuring various isotopes. They showed the same age sequence—young dates near ridges and the oldest dates at the furthest points away. The results resolved some old puzzles. For example, it had previously been known that almost everywhere sediments in the ocean are uniformly less than 200 million years old. Therefore, the ocean floor itself is "young" compared to the age of Earth. It is only a few 100 million years old or less versus a few billion years for some of the continents.

A note of caution is that in Earth science one must be clear about what is meant by the age, because it has a specific meaning for each geological specialty and even for each rock type. There are three principal types of rocks: sedimentary, metamorphic, and igneous. For sediments as in the ocean floor and for sedimentary rocks, the age means when the sediment was deposited. For metamorphic rocks that have been slowly altered from the original rocks, the age is when the alteration took place, and we must accept that the alteration might occur over a long period of time. For igneous rocks, generally from volcanos or lava flows on both the land and ocean and because of isotopic dating methods, the age means the time from when the rock solidified from a molten form.

The detailed and extensive mapping of the magnetic stripes showed that the evolution of the margins is quite complicated in some locations, especially near triple junctions where three plates come together. Even microplates occasionally form, and they sometimes rotate extensively. The plates aren't always small and there is some evidence that Spain has rotated as well. People became so expert at mapping and interpreting magnetic data that they could even identify and name old plates dating back tens of millions of years. Some have been swept into the mantle and vanished.

After arriving in Woods Hole in 1971, I met many of the people who had gone to sea or analyzed the data sets to get all that information. Some worked locally and others were at MIT. We connected because of the MIT-WHOI joint program in oceanography, while others passed through Woods Hole and gave seminars. I found it all remarkably exciting. Some of the reconstructions of plates and their former structure can be found on the web now. A collection for some of the results is freely available from Tanya Atwater at https://animations.geol.ucsb.edu/1_DownloadPage/Download_Page.html.

This is the first half of the story of the discovery of plate tectonics. Since the plates are rigid and all plates are produced at spreading centers, they must be consumed somewhere else. The locations of plate consumption were also discovered in the 1960s, and those made the whole picture even more convincing than the ideas that were promoted by Wegener half a century earlier. The novel phrase, "new global tectonics", was coined by Isaacs, Oliver, and Sykes in 1968 to describe the entire

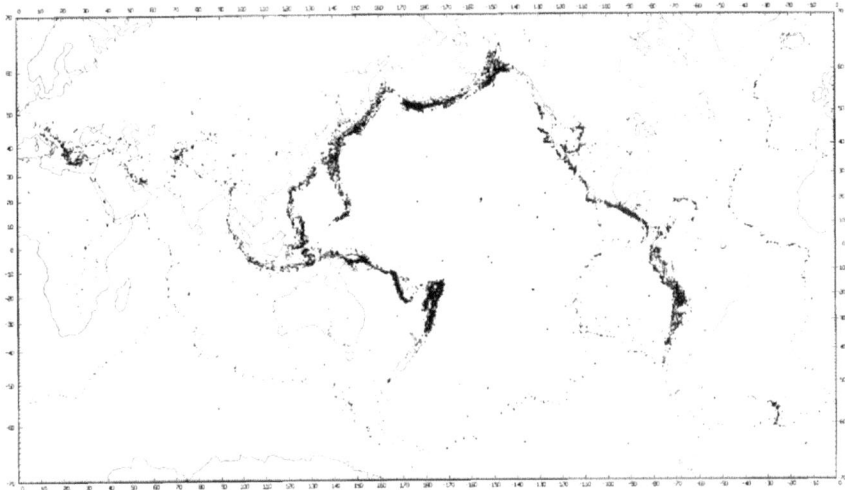

Fig. 1.5 Locations of earthquakes on Earth from 1961 through 1967. The greatest clusters are near subduction zones and less clustering occurs at spreading centers. They outline a few large plates (called blocks at that time). Figure 15 from Isaacs et al. (1968), used under Creative Commons CC-BY license

picture. Although a few years later the phrase was replaced by plate tectonics, their influential study completed the new understanding of plates and three of their figures help to clarify the overall picture. The first of these (Fig. 1.5) shows the locations of earthquakes worldwide. Earthquakes with the most clustering represent sinking locations with the newly named "subduction zones". The spreading centers have lineated clusters. They define the borders of the largest plates.

The earthquakes near subduction zones had long been known to be concentrated and deep. The structure of the "downstream" end of the plates came from advances in seismology that allowed more precise location of earthquakes. They indicated that earthquakes under deep ocean trenches are concentrated in a tilted sheet-like ribbon that extends down hundreds of kilometers. Figure 1.6 shows a typical result showing earthquakes that indicate a brittle slab with approximately a 450 angle under the Tonga arc. Earlier, Sykes (1966) presented many examples of earthquakes within slabs in the Pacific like this.

The concept was put forth and soon verified that those deep earthquakes occurred in cold slabs that bend and sink. Because they are cold and dense, they sink down into the Earth. The deep ocean trenches in the ocean floor above subduction zones are consistent with a dense slab tilting and sinking into the mantle. As in the Titanic movie, the slabs produce a great suction when sinking. Since the ocean sediments do not pile up as they are conveyed to subduction zones, a large percentage of them are swept down into the mantle as well.

This intense movement of the ocean floor indicated a young age of plates compared to Earth's age of 4.5 billion years. The plates are consumed, and ocean sediments are

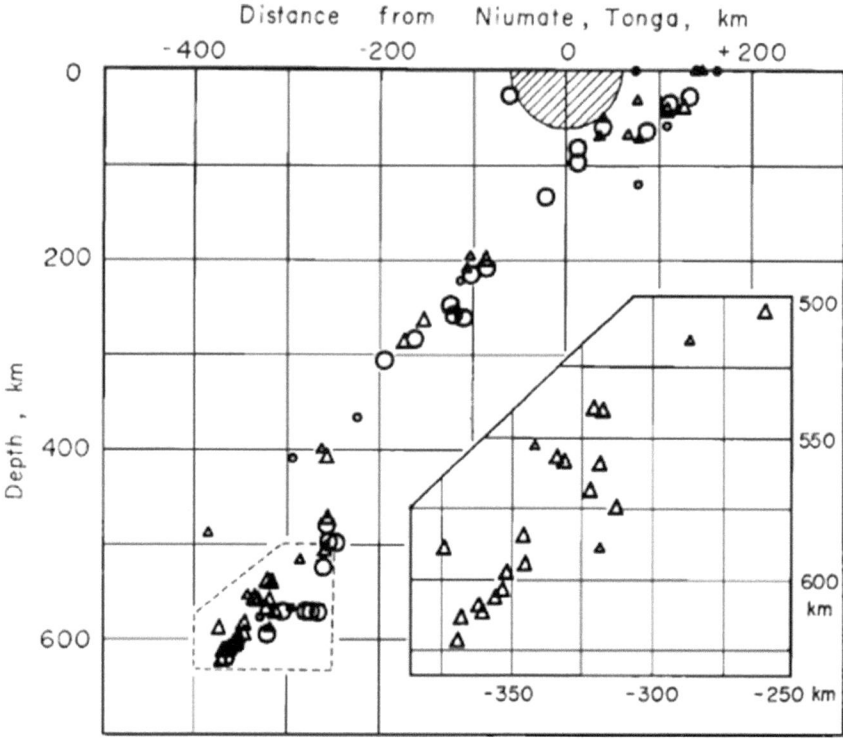

Fig. 1.6 Earthquakes in 1965 were normal to the Tonga arc. Circles represent earthquakes extending up to 150 km north of the section and triangles up to 150 km south. Isaacs et al. (1968) Fig. 9, used under Creative Commons CC-BY license

swept into the Earth or piled up at subduction zones. The study of how plate tectonics occurs at the margins and how they move over the spherical surface is found in the lucid short book by Cox and Hart (1986).

In the next few decades, the rigid nature of the plates and the angles of sinking were used to make many people skeptical that the convection cells that are described in Chap. 2 resemble mantle convection. This skepticism has generally gone away in view of all the other evidence supporting convection in the mantle. Figure 1.7 is a picture that persists to the present. Here are two other useful figures. In Fig. 1.8 the plates and their motions are shown, and Fig. 1.9, is a sketch my current understanding of plate motions along with other important components of motion.

We will now proceed to Chap. 2 which shows how basic fluid mechanics helps us to understand how all the plates and flows are formed. It reviews some essentials of our understanding of cellular convection, including linear and finite-amplitude instability and simple laboratory observations. It is followed by Chap. 3 which brings energy flow into the story by introducing measurements of heat flow of the ocean floor. Mantle material rises, spreads apart, and is cooled under spreading centers.

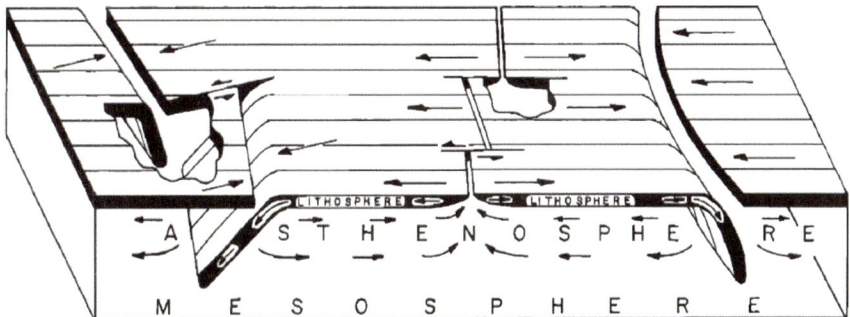

Fig. 1.7 Diagram showing the solid plates of the Earth, with the mantle rising and solidifying at spreading centers with sinking and reheating at subduction zones. Figure 1 from Isaacs et al. (1968), used under Creative Commons CC-BY license

Fig. 1.8 The large plates with arrows showing their speeds with respect to a fixed frame. This projection does not do justice to the plates on a sphere because the plates are curved so the shells slide over the spherical Earth. The fastest speed of the plates is approximately 0.1 m per year or 3.2 * 10^{-9} m/s (from https://commons.wikimedia.org/wiki/File:Tectonic_plates.svg)

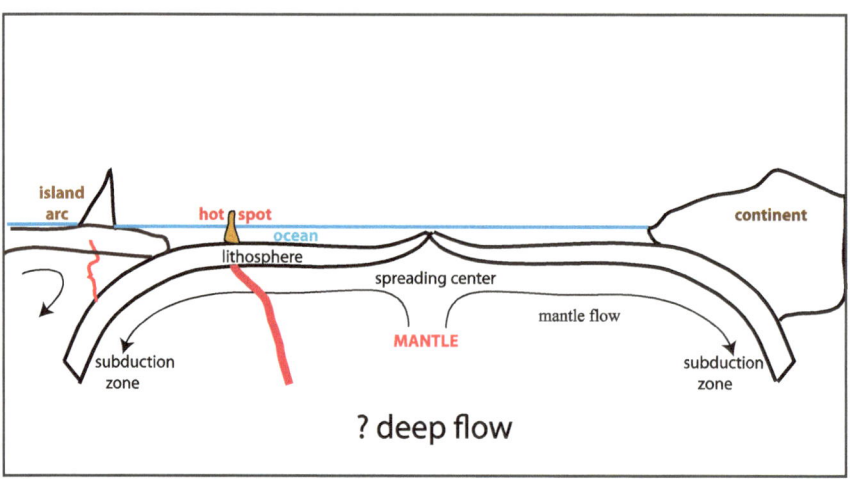

Fig. 1.9 A cross section of the important components of upper mantle motion

This occurs in both cellular convection and Earth's mantle, so the rising and cooling share the same energy-flow pattern.

References

Cox A, Hart RB (1986) Plate tectonics how it works. Blackwell Scientific Publication Ind, Palo Alto

Isaacs B, Oliver J, Sykes LR (1968) Seismology and the new global tectonics. J Geophys Res 73(18):5855–5899

Sykes LR (1966) The seismicity and deep structure of island arcs. J Geophys Res 71(12):2981–3006

Turcotte DL, Oxburgh ER (1967) Finite amplitude convective cells and continental drift. J Fluid Mech 28(1):29–42

Chapter 2
Spontaneous Production of Flow Patterns—Cellular Convection

Abstract The laboratory and theoretical convection cells described in this chapter have energy flow that is very similar to mantle convection which leads to the production of the great tectonic plates. The overall structure of the circulation that exists in a fluid dynamics theory and in the laboratory has dramatic similarities and distinct differences from the mantle. Cellular convection has the advantage that we understand how the cells form and reach some of their final configurations. The first stability study by Rayleigh shows how the flow of heat allows background disturbances to grow if a certain criterion that is named after him, the Rayleigh number, is met. The cells develop a surface boundary layer with high heat flow to the upper surface.

2.1 A Simple Demonstration of Convection Cells

Everybody wonders how things are formed. Mythical stories have always existed about the formation of the Earth, mountains, oceans, and nature. Scientific explanations give some answers, and the next two chapters illustrate a new approach to the ageless questions of Earth, mountains, and oceans. The question is not about atoms or protons or electrons or chemicals, but how they all come together cooperatively to make our world. We begin with a short history of how we came to understand how one very simple thing forms. It is called cellular convection, and it is possibly the simplest energy-flow structure. After learning about cellular convection, how it starts, grows, and matures, and how the flow of energy is involved in all aspects, then we will see that the questions of the formation of land and sea in our great planet (and the magnetic field) are more readily answered.

This example has two processes that I will mention again and again concerning energy flow. The first process is thermal conduction. Heat is a form of energy stored in any material by microscopic random movements, or vibration. Therefore, hot material (for example, water, rock, magma, or mayonnaise) has molecules that are vibrating more than cold material. When hot material is placed next to colder material, the vibrations spread out from hot to cold, and this flow of heat energy is called thermal conduction. The rate of spread is easily calculated. Heat flows from hot material to cold material at a rate depending on the thermal conductivity of the material multiplied by the magnitude of the temperature difference and this product is divided by the distance between hot and cold. Thus, a hot place and a cold place 20 miles apart separated by a material with one value of thermal conductivity might conduct heat at a very small rate, However, two places only one inch apart that are separated by the same thermal conductivity would transport heat at a much higher rate.

The second process is called convection. Convection means taking a hot material and bringing the material to a colder place where it can warm the colder place. Many items in this book possess a form called free convection since a buoyancy force propels them. There are many examples of free convection. For instance, hot stoves have heated air rising around them. The hot air then spreads out to heat the surroundings by free convection. The sun makes black pavement hot in the summer and the rising air heats the atmosphere. Convection is all around us and all around the Earth, too. If the fluid is pumped and not free to circulate buoyantly, it is called forced convection. Forced convection is used frequency by engineers for cooling machinery, and cooling of your hot body by a breeze is a form of forced convection.

Both conduction of heat and free convection can combine to make convection cells. What are these cells? They occur in layers of fluid with the bottom warmer than the top. Lighter fluid near the bottom floats and moves up to the top where the heat is deposited into the top boundary. The top fluid that loses heat to the cold top boundary, becomes denser and sinks to the bottom where it received the heat. The rising and sinking occur as a pattern of vertically circulating flow that can exist as a cellular pattern.

Cellular convection is central to this book. The topic was introduced in the nineteenth century. A publication in 1879 described "tessellated" streaks and swirls in the top surface of hot soapy water. It motivated a PhD thesis by Henri Benard in the period 1890–1900 (Bénard 1901) who produced wonderful photos of careful laboratory experiments with a very flat layer of oil on a hot metal plate that was cooled by the air. The photographs he presented clearly showed hexagonal cells everywhere. Photos are now easily found on the web. The assumption was that the cells were a form of convection with the vertical circulation described above. Although the photographs are outstanding, I show here a photo of a simple experiment in my kitchen that you can do, too. The kitchen experiments are better (in my opinion) because you can do them yourself! They can be repeated many times, and you can move your head around to see the three-dimensional nature of the cells. You can see the cellular shapes form and then later mix away and of course you are producing them yourselves.

The experiment is simple. You just need to take a frying pan with a dark flat bottom and place it on a cold level stove burner. First, place about 1 cm (1/2″ or so) of water at room temperature or warmer in the frying pan. Leave it on the stove for at least a couple of minutes so the water circulation slows down. Second, take some cold milk from the refrigerator and using a cup or glass, (about 1/2 cup is enough) very gently dribble the cold milk in along the inside edge of the pan. It helps to have the tiny stream of milk touch the pan and avoid turbulence. A small bit of paper towel draped along one edge also helps by allowing the poured milk to fall onto the paper-water intersection with little mixing. The cold milk is denser than the warm water, so if there is not too much mixing, it sinks below the water and spreads out over the bottom making a bottom layer of milk about 1/2 cm deep (1/4 in.). Waiting a minute or two will allow all the flows to slow down. After everything is still, turn on the stove heat. If you are using a gas stove, turn it to low. Soon (between 10 s and a minute later), you will see the energy-flow structures. These are convection cells in the milk. The suddenly warmed milk at the very bottom of the liquid layer floats up into the water above it with the form of a cellular honeycomb of blobs and sheets. The clear colder water slowly moves down to help fill in around the rising, light, newly warmed milk. Some cells that I photographed on my stove are in Fig. 2.1.

What are the cells composed of and how do they arise? Nobody knew until a mathematical explanation was produced by Rayleigh (1916). He calculated what we call "an instability" of the layer of motionless fluid. What do I mean by "instability" of the layer? First, we know that in our stove experiment when the stove is turned on, the pan gets hot. The heat moves up through the metal bottom by thermal conduction (Metal has a high value of thermal conductivity and quickly gets hot as we all know when we touch a hot piece of metal). The fluid right above the metal bottom gets hot next. The thickness of the layer of heated fluid is at the first instant zero and in a couple of seconds it has a very tiny thickness, less than 1 mm or 1/16th of an inch. The milk sits above the bottom without any appreciable motion. Then, the heated

Fig. 2.1 A photograph of convection cells forming in a thin layer of cold milk lying under a layer of warmer water in a frying pan on a stove after about 20 s of heat was suddenly turned on (Photo taken by the author)

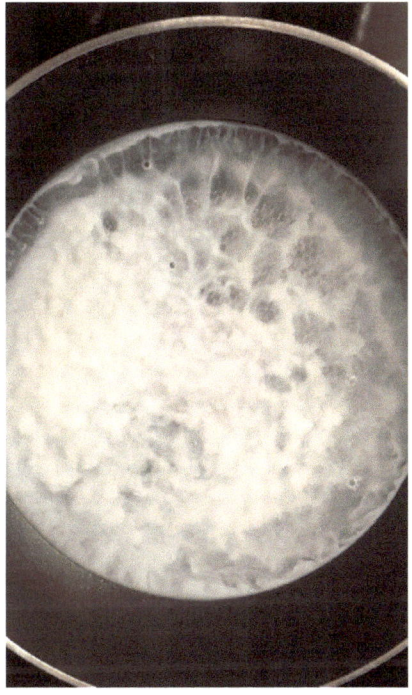

Fig. 2.2 Rayleigh's
perturbation cell

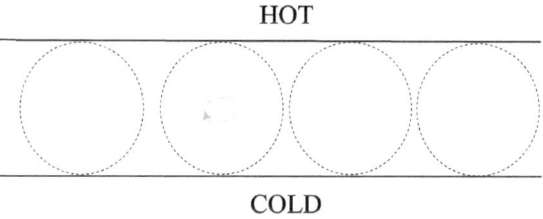

HOT

COLD

fluid thickness increases and instability occurs when the hot milk begins to float upward around colder water.

Rayleigh's theory (1916) about how this flow develops is remarkable. He uses calculus equations that are very precise for a heated fluid. They exactly duplicate the flow of a liquid and the flow of heat everywhere. I worked through his equations as a graduate student. He began with a very simplified situation with the temperature evenly changing from hot at the bottom surface of a fluid layer to cool at the top surface in a field of gravity. There is no temperature change in the two lateral directions. Then, he used equations for a very tiny disturbance (the technical word we use is "perturbation") and found a formula telling the speed that the perturbation either grows or decays. What drives the perturbation so that it grows? As we have mentioned for convection cells, upward flow brings hot fluid upward, and downward flow brings cold fluid downward. The lower density of the rising fluid and higher density of the sinking fluid produce a buoyancy force that drives the motion. But why are there cells appearing in the motionless fluid? Where did they come from? How big are they? The equations showed that any perturbation can be the sum of circulating cells of many different sizes (one size is shown as dashed circles in the sketch in Fig. 2.2, but in general, any random background noise from tiny molecular fluctuations makes thousands of cells of different sizes). Moreover, his equations were such that we can look at the properties of each cell with a specified size alone. Rayleigh calculated the rate of growth or decay with time of the speed of each cell with each size. It is important to emphasize that the cellular nature and Rayleigh's ability to analyze each cell individually are mathematical facts rather than physical facts, although they are based on equations for a fluid, which are physical facts. His calculations showed that a cell extending from top to bottom grows faster than one with many stacked vertical cells, and he also produced a formula for how quickly a cell of a given wavelength in the two lateral directions grows or decays in time. Let us be clear here! The mathematics is needed, but the physics in the stability problem are represented by the formulas that Rayleigh's produced. These formulas include many components of the physical fluid. Three distances are involved, the depth of the fluid layer and the wavelengths of a cell in two lateral horizontal directions. They also include some physical properties of the fluid: temperature difference between top and bottom boundaries, acceleration of gravity, fluid viscosity, fluid thermal conductivity, fluid coefficient of thermal expansion, fluid density, and the fluid specific heat (i.e., the amount of heat needed to change a gram of the material by one-degree centigrade). We also must be clear that his equations were limited to predicting the quickness or the growth rate of the tiny flow. Specifically, each wavelength, and the accompanying internal cell structure, either grows or decays in time starting from the smallest random background disturbances, no matter how small they are. The mathematical solution showed that one wavelength grows most quickly, and many different cell arrangements have the same value of wavelength. This was an exciting hint about the cellular nature of Benard's flow. The cells with the size within a range of wavelengths grow larger with time and others become smaller. Many cell patterns, such as rolls, squares, hexagons, and so forth have the same wavelength in this theory. If the lateral size of a chamber becomes much greater than fluid depth, the size is much less important than the inherent range of cell sizes that grow!

2.2 Rayleigh's Number and Linear Stability

Since Rayleigh found the formula, people named it "the Rayleigh number" in his honor. The formula for the Rayleigh number is denoted by the symbol Ra and it is worth displaying.

$$\mathrm{Ra} = \frac{g \times \alpha \times \Delta T \times d^3}{\kappa \times \nu}.$$

The symbol for multiplication ($\times$) is used in this book to avoid confusion. The other symbols are g for the strength of gravity; α for the coefficient of thermal expansion; ΔT for temperature difference; d for the depth of the layer, κ (kappa), for thermal diffusivity, which expresses the rate of temperature spreading out by thermal conduction; and ν (nu), for the kinematic viscosity, which expresses the rate of stress spreading out in a fluid.

Rayleigh number is full of physical meaning. For example, it contains two terms expressing speeds and they are divided by each other. One term is the speed of flow as in Chap. 1 and the other is the speed at which thermal conduction would heat up the entire fluid if motion is absent. It also expresses two terms expressing acceleration. One is the gravitational acceleration of a hot rising motion or cold sinking motion, and the other is the deceleration by viscous drag. Finally, Rayleigh number applies to not only this specialized problem but much more generally in the Earth as we shall see.

Rayleigh's theory produced a value of Rayleigh number that separates growing perturbations from decaying ones. This value is called a critical Rayleigh number, and it is represented here by the symbol Rac. It might have one of several different values, depending on the nature of the top and bottom boundaries. The first stability study applied to fluid lying in a layer with two constant values of temperature above and below and with surfaces that allow the fluid to freely slip. It found that $\mathrm{Ra}c = 27\pi^4/4$ or approximately 657. This value is not sacred. Later studies found that other values for Rac depend on whether the top and bottom boundaries were either slippery or stiff and whether either temperature or heat flow is imposed on the surfaces. In addition, another value of Rac can be found for fluid that is internally heated or cooled, but in that case, a heat production version of Ra must be defined. In this example that we have described so far, (cells with constant values of the physical parameters) if Ra is less than Rac, the motionless layer is stable and all the perturbations decay away in time. If Ra is equal to Rac, cells with one value of wavelength stay steady as in Fig. 2.2, and all other wavelengths decay away in time. Above Rac, cells with a small range of wavelengths grow in time and other wavelengths decay.

2.3 Dimensionless Numbers and Scaling

What does all of this have to do with Earth's giant engines? The answer is deceptively modest but extremely powerful and important. First, the Rayleigh number formula has no dimensions! It has no length, no age, no force, and no mass. This means that if any situation has an appropriate Rayleigh number, then Rayleigh's analysis can be applied. Since the layer depth is cubed in the formula for Ra, there is a clear hint about the size of layers where instability can happen. For example, layers of greater depth than laboratory values require lower values of $g \times \alpha \times \Delta T$ or higher values of $\kappa \times \nu$ to be at critical value. More importantly, it was recognized in Rayleigh's era that dimensionless numbers govern all sorts of fluid flow and heat transfer behavior, not just stability. Dimensionless numbers are the basis of all kinds of forecasting projects that we depend upon all the time now such as weather prediction, climate prediction, solar cycles, rainfall, and so forth. For example, we can conduct laboratory experiments and numerical model calculations with a Rayleigh number like the inside of the Earth. The same is true in numerous areas, for example for convection cells in the sun, or in the core of the Earth.

As a historical note, other studies in that era (1880–1920) found similar instabilities for flows governed by other dimensionless numbers. Results included the stability of shear flow within a flowing liquid, the growth of vortices in swirling flow, the production of sound in compressible flow, and a moving liquid that conducts electricity. In some cases, growth calculations were so difficult that it was easier to find a mathematical proof that something is impossible. For example, it was proven that flows cannot become turbulent if a dimensionless number called the Reynolds number (velocity times layer depth divided by kinematic viscosity) is below a given value. It was also proven that a two-dimensional flow of an electrically conducting fluid cannot develop a perturbation that produces a magnetic field no matter how high the speed is.

Overall, the strength of theories like the one developed by Rayleigh and by others is that they help us to know important ranges of a dimensionless number for something to happen. Another strength is that the mathematics can be confirmed by others as I can attest, first as a graduate student and then throughout my life. This clarity makes the results clear to everyone. A third strength is that the prediction lasts forever. I even have questions about whether to use past or present tense in reviewing previous work since the correct math and the physical basis are unchanging. A fourth strength is that the results occur with less expense than any experiment could hope to do. Some predictions were experimentally verified with great success but in other cases, experiments are not possible. There are weaknesses, of course. A glaring weakness of the early theories is that nothing could be calculated after the tiny perturbations grow bigger. Also, they predict the growth of many different cellular patterns and not just the hexagons that Benard reported.

2.4 Instability—The Growth of Practically Nothing

This reviews how a perturbation grows as predicted by Rayleigh and serves as an example of dimensionless scaling. Our example is a numerical computation. It accomplishes our task of following how the flow moves ahead in time while various wavelengths of the perturbations grow or decay. It uses a method called finite difference, which is a common general approach. The numerical code used advances in small time steps using the calculus equations that express the conservation of momentum, mass, and heat for two-dimensional fluid flow, so velocity in one direction is always zero. The equations are applied within a rectangular cross-sectional chamber with a higher temperature on the bottom. It starts with exactly the problem described by Rayleigh (a motionless fluid with a temperature change that smoothly changes from high temperature along the bottom boundary to lower temperature along the top). It is also like the frying pan experiment in Fig. 2.1. The motionless fluid has thermal conduction of heat upward. We can calculate the value of energy flowing upward with a simple formula. The temperature difference is multiplied by thermal conductivity and divided by the depth of the layer. Then, a tiny perturbation is added to this energy flowing upward.

Figures 2.3, 2.4, 2.5, 2.6, 2.7 and 2.8 show the results of a numerical calculation. The entire code and therefore, the output uses dimensionless units. What does this mean? First, the temperature everywhere in the numerical computation (and in the figures) has been changed from the usual units of temperature. The average temperature of the top and bottom boundaries is subtracted from the fluid temperature and then it is divided by the temperature difference between bottom and top ΔT. This makes our temperature within the computer code dimensionless, and the difference in value from bottom to top is equal to 1. In the numerical text of the code, the dimensionless value of temperature is 0.5 at the bottom and -0.5 at the top. Second, all lengths are dimensionless, too, because they are divided by the real layer depth which ordinarily would be in units of meters. Third, the time is divided by a time it takes for thermal conduction to penetrate up and down and change temperature in a motionless fluid. The formula is called a thermal timescale, and it is d^2/κ (This expresses in seconds how quickly the layer can heat up). Therefore, the time in the figures is dimensionless. Fourth, the speed has been divided by a velocity scale that is defined by the real depth divided by the timescale. Finally, the calculations are for a very viscous fluid such as a thick oil, a cold syrup, or even a large body, such as the Earth's mantle or glacier ice. Inertia is completely absent in the numerical code, so this flow is different from the movement from classical physics of a ball dropping from your hand down toward the ground with little friction and almost pure acceleration. In fluid mechanics, this exists if a dimensionless number called the Prandtl number Pr is very large, mathematically in the limit $Pr = \infty$.

Computers are great in some ways. I can start with any perturbation I desire. The numerical grid has 32 vertical locations and 256 lateral locations. Values of velocity and temperature are stored at each of the 32×256 grid points and then the calculation advances in time. Technically, the dimensionless time step is 10^{-4} but other values will give approximately the same results if the time step is not too different from this. The value of Rayleigh number is Ra $= 1000$, which is approximately 50% above the instability value of Rac $= 657$ predicted by Rayleigh for a layer that is unconfined laterally. Before the first step of the calculation, the velocity is zero everywhere, and the dimensionless temperature at each grid point in the numerical calculation changes uniformly from bottom to top except that it also has a perturbation that randomly scatters a perturbation temperature at every grid point that is of magnitude 10^{-11}.

We will follow the growth of the instability by mapping out the velocity and temperature fields at different times. Figure 2.3 shows results that are generated after the very first step of this numerical calculation (Another strength of the computer is that all the properties can be determined everywhere at every time step). The top panel shows that the imposed random temperature field has immediately produced a velocity field (also random) that is extremely small with speeds below the dimensionless value of 3×10^{-13}. This type of expression of very small or very large numbers

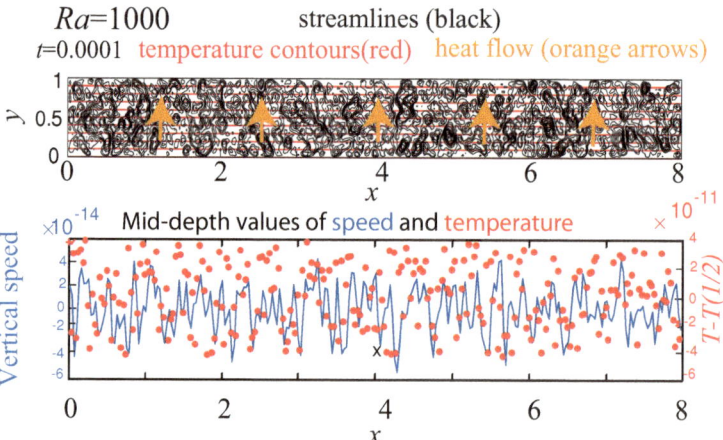

Fig. 2.3 The flow fields and temperatures of a fluid after the first step in time at $t = 0.0001$. The closed streamlines in the top panel represent the circulation that is produced directly by the random temperature perturbations. The red lines are isotherms of the total temperature that is orders of magnitude larger than the perturbations. The large orange arrows show the uniform flow of heat upward by thermal conduction at this initial time. The lower panel has the values of vertical speed and of the temperature of the random perturbation field exactly along the horizontal grid row at mid-depth

with factors of ten is explained in Chap. 9. Streamline contours are shown in the figures. They can be converted to values of speed and the calculation is simple. we divide the stream function value by a distance that makes it zero. The speeds are much too small to affect the flow of heat that is shown by the orange arrows. Heat flow is almost entirely determined by the thermal conduction of heat. The temperature field is also shown by red lines. The bottom panel shows the random temperature and the calculated speed profiles at the grid points lying along the mid-depth of the cell.

Ten calculation steps later, at $t = 0.001$ (Fig. 2.4), the random eddies are smoother and larger. They begin to sweep heat around, changing the temperature and velocity fields. Temperature at mid-depth is shown in the bottom panel and velocity is indicated in both panels. Their fields are a bit smoother and begin to adopt similar shapes. The speeds are still so small that the motion sweeps heat at a rate that is much smaller than the conduction of heat, so the flow of heat remains conductive and unaltered overall. Therefore, the isotherms in the top panel are still straight horizontal lines. The vertical speed has increased by almost a factor of 10, but the temperature amplitude is smaller, probably from smoothing. The smoothing of both temperature and velocity perturbations is consistent with a small range of wavelengths that have grown in time and shorter and longer wavelengths that have decayed. The theory has the fastest growth for wavelengths that are approximately the depth of the chamber.

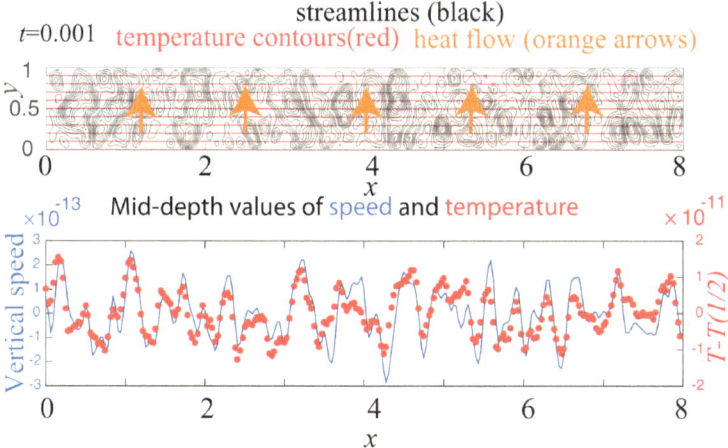

Fig. 2.4 The flow field and speeds of a fluid at $t = 0.001$. As before, the closed streamlines in the top panel represent the circulation and red lines are constant temperature values. The large orange arrows show the uniform flow of heat upward by thermal conduction. The lower panel shows the values of vertical speed and temperature of the perturbation field at mid-depth

One hundred calculation steps after starting at $t = 0.01$ (Fig. 2.5) both the temperature and velocity profiles have become smoother, and the shapes of speed and temperature in the lower panel are closer to being the same as each other. Velocity has grown by almost a factor of 10 but the peak temperature of the perturbation has become a bit smaller, possibly from diffusive smoothing. The magnitudes of both velocity and temperature perturbation values are still extremely small (10^{-12}).

When $t = 1$ (Fig. 2.6), the numerical calculation has been repeated 10,000 times and the pattern of flow seems to be approaching equilibrium. The profiles of vertical

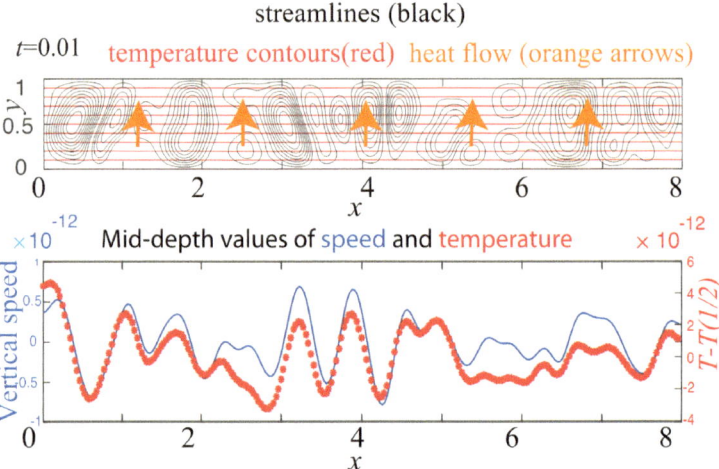

Fig. 2.5 As in Fig. 2.4 at $t = 0.01$

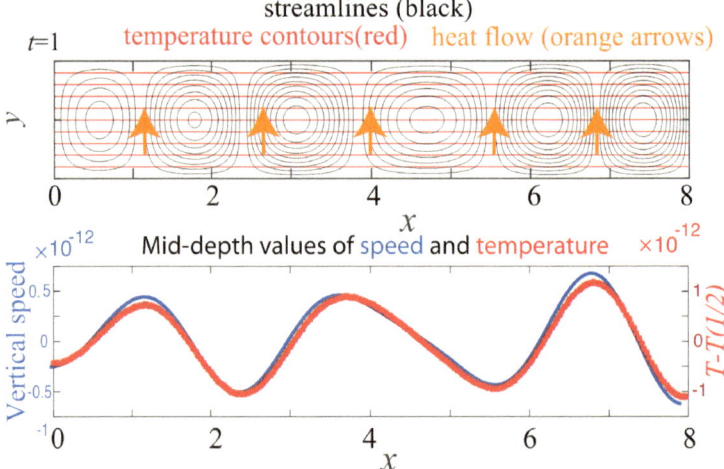

Fig. 2.6 As in Fig. 2.4 at $t = 1$ except that the vertical coordinate for the top panel is stretched by 50% to give a clearer picture of the patterns

speed and temperature at mid-depth are almost identical and have become much smoother than they were before. This shows how dramatically the pattern selection works to make a cellular pattern. The flow has not finished growing because convection is still so small that thermal conduction dominates vertical energy flow with straight horizontal red isotherms in the top panel. Speed and temperature perturbation amplitudes are still small (10^{-12}).

Ten times later (Fig. 2.7), the temperature and velocity values are close to the final ones. Vertical speed and temperature speeds approach a value of order one, and the two profiles have virtually identical shapes. The large-scale temperature field that was shown by red horizontal lines at previous times, is now strongly distorted by the vertical motion of the convection cells. This flow is a perfect example of the flow called cellular convection. Rising fluid is warm and sinking fluid is cold. The convection has changed the flow of energy upward as shown schematically by the uneven orange arrows. The heat flow propels the convective circulation, with the rising hot fluid and the descending cold fluid releasing potential energy at a steady rate. The heat flow is greater in magnitude overall (shown by orange arrows) and is more concentrated at both the top (at the upwelling locations) and at the bottom (at the downwelling location). Finally, the convective energy (heat) flow also takes two forms. One form is convection of heat from upward flow from the bottom hot to the top colder region, and from downward flow from the top cold to bottom hotter region. The other is by thermal conduction close to the boundaries. These convection cells are therefore our first example of an energy-flow structure. They arise from an instability process that selects the shape and eddy wavelength that, when fully matured, alter the heat flow.

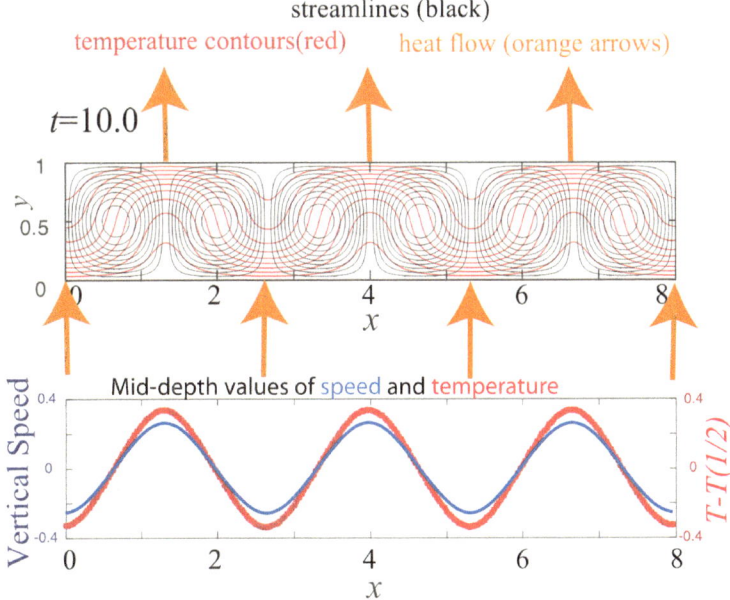

Fig. 2.7 As in Fig. 2.6 at $t = 10$

Convective flows have been extensively studied over a wide range of Rayleigh and Prandtl numbers. A central feature is the formation of top and bottom boundary layers that are clearer in the same calculation for steady flow with a larger Rayleigh number of Ra = 10^4 (Fig. 2.8). The convection distorts the temperature field even more to shapes like plumes. As before, the vertical motion consists of rising and sinking plumes of hot and cold fluid. The top and bottom of the thermal boundary layers become smaller as Ra increases.

This calculation shows how the buoyancy forces combined with energy flow produce this cellular flow. The cells spontaneously grow with flow taking a form that releases buoyancy and increases the rate of release of potential energy. The cells themselves distort the temperature field to make the form of the flow. The cells circulate hot fluid up and cold fluid down, and they allow the flow of potential energy to grow. The flow is also associated with the flow of heat upward and the value becomes greater than the value from thermal conduction. Let's forget the science for a moment! The energy flow works within the material like an invisible ghost, and the forces make the skeleton of the cells.

Focusing once more on energy-flow structures, remember that the energy flow in a motionless fluid takes the form of thermal conduction of heat moving upward from the bottom surface to the top one. Since a field of gravity exists, we can picture that the flow of heat upward is cooling any material below the layer, making it denser. This lowers the center of gravity and releases potential energy. Heating material above the layer makes it less dense and expands upward, raising the center of gravity. Therefore,

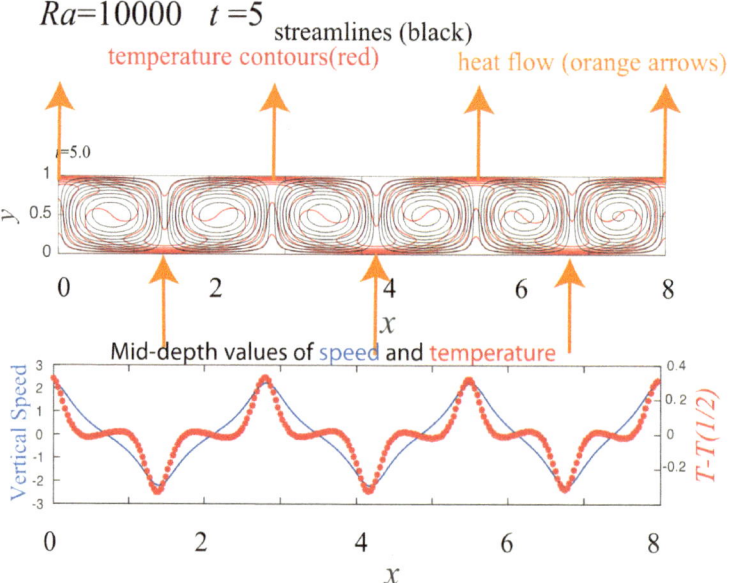

Fig. 2.8 A steady flow at the larger value Ra = 10^4

potential energy flows upward at a steady rate and the speed is proportional to the heat flow rate. Be careful to not become confused between the flow rate of gravitational potential energy and thermal energy. Heat flow upward is also the flow of thermal energy and it has a different magnitude than the flow of potential energy.

In summary, if the Rayleigh number is small, all perturbations decay and there are no growing cells of any wavelength. With a larger Rayleigh number, tiny perturbations with wavelengths in a certain range grow. They begin to use the flow of potential energy to form a velocity pattern that moves the energy up more rapidly than thermal conduction alone. As the perturbation grows, that new velocity field increases, and energy flow increases with it.

The exact numerical value of the critical Rayleigh number is not crucial to this picture. It depends on some details that don't concern us here, for example: whether the boundaries are solid, whether they allow some lateral slip, or whether the temperatures at the bottom or top boundaries remain constant. For example, with solid nonslip boundaries at completely fixed temperatures above and below the fluid, Jeffreys (1926) initiated studies showing that the critical Rayleigh number is slightly above the value Rac = 1707. In contrast, Rayleigh's original configuration has Rac = 657 because the top and bottom surfaces allow the fluid to slide along them.

In this example, the flow is strictly two dimensional for simplicity. We have learned nothing about the pattern in the two lateral directions. Strictly speaking, in his analysis, many different patterns in the two lateral directions are possible: rectangles, squares, hexagons, and rolls. Mathematically, the wavelengths that define each pattern in the two directions can be combined to define one number called a wavenumber, which is inversely proportional to various combinations of the wavelengths of the cells in two directions. Rayleigh's original theory and later embellishments found that all patterns of the "perturbation" (the small growing velocity field) with the same wavenumber have the same instability growth rate. In the actual selection process for the optimal wavenumber only one wavenumber, called the critical wavenumber, has zero growth at Rac all others decay in time. It is most able to liberate buoyancy force to balance drag and the spreading out of the temperature perturbation. Slightly above the critical Rayleigh number, cells with a small range of wavenumbers (a bandwidth) grow in time, and all wavenumbers outside this range decay in time. This is one of the simplest examples of self-organization I know. Even so, the critical wavenumber represents many possible three-dimensional patterns.

Ironically, after all this work, the cells seen by Benard were discovered to be driven by something different from the buoyancy mechanism described here. The surface tension variations of the free oil surface, called the Marangoni effect, rather than the buoyancy driving, which was included in the theory by Rayleigh, drive Benard's cells in the laboratory. This is a good example of the stimulus of a useful theory by an incorrectly interpreted experimental result.

Meanwhile, laboratory measurements were used to determine how heat and momentum moved around after the fluid became unstable. Convection cells driven by buoyancy were readily made for layers of liquid or gas confined between two solid surfaces. Generally, the solid surface material was either metal or glass. The circulating convection cells were observed to have a roll rather than a polygonal structure and to exist over a considerable range of Ra above Rac. No theory before the mid-1950s could show how this happens. Above the range with rolls, the flow became irregular, (Getling and Brausch 2003) and no theory could tell how turbulence mixes everything up.

Summarizing, numerous stability studies for the growth of tiny perturbations were published but two important questions remained unanswered. First, although the unstable perturbations grow exponentially, ultimately the growth must stop and at that time the question of, "how is equilibrium achieved?" remained with no answer from theory. Second, what are the selection principles for the patterns (hexagons, squares, or roll cells)? Until the mid 1950s (a few years before I was a graduate student) scientists were not successful in calculating answers to either question, both for cellular convection and for a wide range of other physical problems. The final "sizes of things", as well as certain patterns of the growing "things", could not be calculated. Then, theoretical work that became to be called finite-amplitude theory came along. It introduced understanding that paved the way to application in the natural environment. That is the next step in our story.

2.5   Finite Amplitude Theory

At this point, we must remember that computer calculations like those shown in Figs. 2.3, 2.4, 2.5, 2.6, 2.7 and 2.8 onward were not available until digital computers were made available for general scientific use in the mid-1960s, and calculations were limited to stability theory for the growth of tiny perturbations. Once computers entered the picture, everything changed. Calculations became so easy to make thereafter that many universities did not allow Ph.D. dissertations for Rayleigh's type of stability theory. Understanding and calculating how the finite-amplitude cells grow to a steady wavelength and what governs their structure was given a boost from laboratory experiments in 1954 by the physicist Willem V. R. Malkus. He observed in cellular convection experiments that progressive changes in the slopes of heat flow measurements occur as the Rayleigh number increases from Rac to much greater values. He plotted the results using logarithms.

(Logarithms are used by scientists for first looks at data because they indicate whether there is a power law relation between two variables. For example, consider the relation $x = yb$. Its logarithmic expression is $\log x = b \times \log y$, which on a graph is a straight line with slope b.)

The physical interpretation was that at progressively greater values of Ra, a change of the slope implies that the first cells have grown and reached some steady flow like those in Figs. 2.7 or 2.8, some smaller (in the vertical direction) internal cells will grow. These will produce both greater heat flow and a greater rate of potential energy release. This idea proved to be a useful guide for predicting the equilibrium speed and the heat flow of the developed flow. Based on his experiments, Malkus invoked in a separate paper a mathematical model of how the maximization of internal heat transport drives these internal cells. At that time many others such as the famed

Russian Physicist Lev Landau and Belgian chemist Ilya Prigogine had been concurrently suggesting similar ideas without specific experimental results to compare with simplified theories. To this day, such ideas have received great attention in the physics community. Ironically, the concepts that a maximum transport of either heat flow or anything else determines entirely a flow structure, although stimulating, have never been found to be 100% true.

The next step was a successful complete theory that began to finish Rayleigh's pioneering development. It is a classic story of the teaming up between a physicist (Willem V. R. Malkus), and an applied mathematician (George Veronis). They developed a method to calculate the speed that the flow achieves when it comes to equilibrium and brings the growing fields of temperature and velocity to a steady balance. The flow of heat upward to increase the rate of potential energy flow played a large part. In particular, the flow of heat changed the vertical temperature profile as the motion took over the role of heat flow as Fig. 2.8 showed. The successful analysis of convection and the clarity of their approach make their paper "Malkus and Veronis (1958)" stand out. Historically, I can tell you that although Malkus was listed as the first author, I was told by a reliable source that most of the mathematics and writing of the paper was done by George Veronis, who had decided early in his career to insist that all authorship of his papers should be listed alphabetically. Without question, all of us who knew Willem knew that his prodding as a physicist was significant, but the true kudos for the actual mathematics go to George Veronis! Other mathematicians (literally, from around the world) were making similar developments almost simultaneously and Malkus and Veronis gave them full recognition.

Their result was a significant breakthrough and a dramatic success. They successfully started a well-known summer school that helped everyone work on these and similar studies for decades (Fig. 2.9). The technique was the first to make the prediction of the speed of the fluid, but the issue of the form in the two lateral directions of the convection cells was not completely clarified as each pattern had different calculated speeds. Next year, Palm (1960) extended the technique to include hexagons and he concluded that a variation in viscosity with temperature was needed to trigger the planform. Later, it was found that temperature dependence for other physical properties, such as thermal diffusivity or the coefficient of thermal expansion, produce similar results. From roughly 1959–1968, legions of mathematicians and theoretical physicists used not only the "Malkus-Veronis" original approach, but they also quickly developed modifications to unload an entire blizzard of calculations to the world. The techniques are generally called "finite-amplitude analysis".

Applications were made to numerous branches of engineering, applied mathematics, physics, and Earth science. Many new structures were analyzed and some of the solutions varied in time to include explosive instabilities, chaotic motions, and great alterations to the basic flow. The community also produced proofs of absolute stability below critical values and more. I would dearly love to name all the wonderful contributors and their work here, but it requires books and in fact, many already exist.

Focusing back on convection cells and energy-flow structures much was learned using finite-amplitude analysis. It was found that the hexagon pattern is somewhat special since it occurs only if viscosity, conductivity, or thermal expansion depends on

Fig. 2.9 Willem Malkus (right) and George Veronis (left) on the porch of the fabled Walsh Cottage. They established the highly successful Geophysical Fluid Dynamics Program in 1959 that is held for ten weeks each summer. The program continues up to the present. Not only have I continued to attend since 1972, but I also learned much of the material in this book during lectures there (Photo taken by the author)

temperature or with certain conditions along the top or bottom boundaries. Benard's original experimental flows are not even buoyancy driven but driven by a temperature dependence of surface tension on the oil surface exposed to air. Theory produced other guidance too. A certain time dependence was equivalent to internal heating, and both can also produce hexagons.

This brings me to work in the mid-1960s when I was eagerly watching as a graduate student. The theory showed that hexagons might appear just after an experiment was started and vanish as a steady state was reached. Let us see why this is. The Malkus-Veronis approach predicts that the speed of flow increases for progressively larger values of Ra above Rac. Many different values of wavelength are possible but all of them are two-dimensional roll cells as in Figs. 2.6, 2.7 and 2.8. Hexagons occupy an additional branch in the approach because with variation of material properties they liberate potential energy flow better than rolls in a small range close to Rac. However, to exist there must be modification to the basic states such as internal heat production, a slow time variation of the bottom or top temperatures, or a variation of one of the physical properties with temperature. In addition, hexagons occur best only for a finite range in Ra, and beyond that range the pattern consists of a single set of rolls that are strictly two-dimensional. Ruby Krishnamurti, a student of Malkus, conducted pioneering work in that direction. Naturally, in a layer of material, the roll axis can lie in any lateral direction.

Overall, the theoretical predictions for the stability of cellular flow were a great success because not only did they show the conditions needed for growth, but the form, stability, and the cellular nature of the evolving flow were found. The fact that Rayleigh predicted a range of cell wavelengths that can grow if Rac is exceeded introduces some interesting features. I contributed to some answers both with theory and experiments. In a graduate student project, I found a way to initiate rolls with strong light shining through a grid. The rolls thereby had a fixed wavelength when Ra is below Rac, and then we could observe their stability after Ra exceeds Rac (Chen and Whitehead 1968). Then, in the Malkus laboratory at UCLA, we observed in the laboratory the zig-zag and cross-roll instabilities that were predicted for roll cells by Fritz Busse (Fig. 2.10) (Busse and Whitehead 1971). In addition, I helped with some theory about how cells in a chamber of a given length adjust to the cell size (Newell and Whitehead 1969), and how roll cells develop the instability in the third direction. I was very lucky. They were wonderful times!

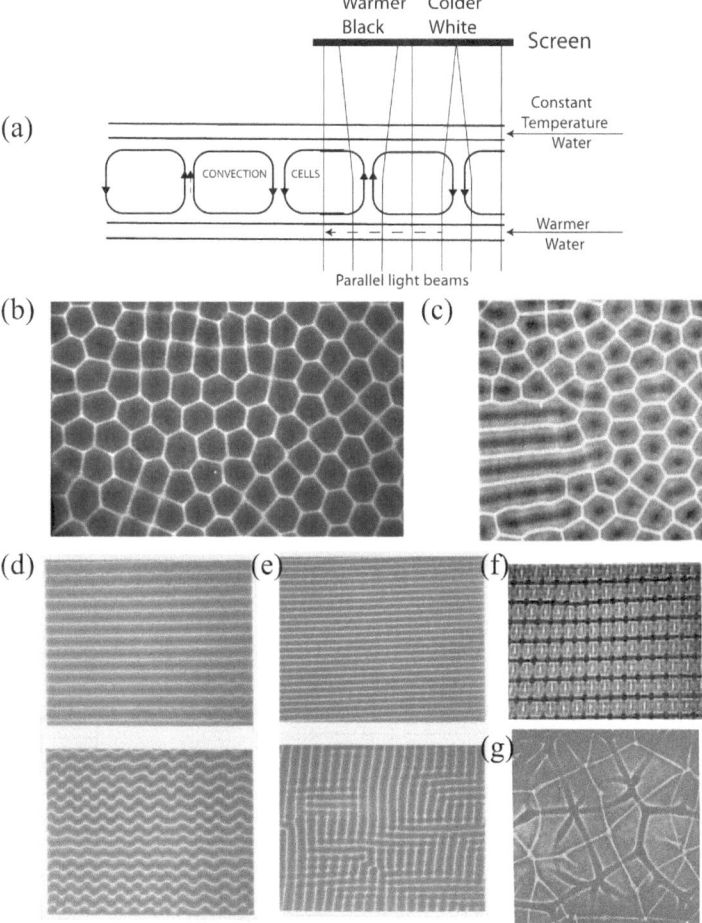

Fig. 2.10 **a** How a shadowgraph focuses light. Cold sinking fluid is lighter, and warm rising fluid is darker. **b** Top row, hexagons at Ra = 1750 and a transition to roll cells **c** at Ra = 1800 (photos taken with B. Parsons). **d** Large wavelength roll cells at Ra = 3600 with the zig-zag instability are shown below. **e** Small wavelength roll cells at Ra = 3000 with the cross-roll instability shown below. **f** Bimodal flow at approximately Ra = 50,000 (Busse and Whitehead 1971). **g** Spoke convection at Ra = 100,000 (Whitehead 2003)

Results were extended to other regions of parameter space or other configurations in various studies. New patterns emerged and time dependence occurred. In general, higher Rayleigh number flows have multiple cells in different directions, and fluids that have small viscosity (compared to thermal diffusivity values) have energetic time dependence and more complicated patterns. Convection in stars is a good example. New patterns were also discovered for convection cells in spherical shells, in cylindrical flows, and in fluids of more complicated properties. The patterns and speed of cellular convection became well-studied theoretically around the mid-1960s and I was fascinated. Busse's (1978) review article "Non-linear properties of Thermal Convection" is the best overview of the remarkable progress that starts with Malkus-Veronis type of calculation and then finds successive new cellular flows at greater Ra. He developed them so they could be extended into a large range of parameters and his results have been extended to spheres, spinning cylinders, turbulence, and shear instability. He also extended the methods to convection in electrically conducting fluids, where fundamental aspects were discovered about flow producing a magnetic field (see Chap. 6). I feel very fortunate to have worked with this gifted scientist.

Jumping back in time for a bit, those newly developed theoretical predictions were big news during the 1960s when I was a graduate student looking for a thesis topic. Many different laboratory experiments were being developed to show the behavior of these convection cells. I frequently saw hexagons (as in our Fig. 2.1) in experiments that usually went away when the experiment settled down to steady roll cell structures. To conduct good experiments, scientists used very flat horizontal layers and very carefully controlled the sidewall conditions. Naturally the experiments could have many different starting conditions. Some experiments were designed to see the flows for hexagons in convection, rather than the original surface tension driven flows. In my thesis the fluid was heated from above by very intense light, and the flow developed circular sinking spouts of cold fluid with warm fluid rising to the top everywhere else. They correspond to hexagons with sinking at the center. Figure 2.10 shows how a shadow graph concentrates light to reveal a pattern on a screen. It also has some shadowgraph photographs of patterns in some of my experiments. These include hexagons, rolls that become unstable to other rolls, two sets of rolls at right angles that we named bimodal convection, and a spoke pattern at even higher values of Ra.

One can imagine that each new mode of flow liberates a greater degree of freedom with increasing Ra. Figure 2.11 shows a comparison of calculations by Busse and an experiment for cells generated by a light grid after the induced cells with the wavenumber, defined as 2π/wavelength, reached the shown Rayleigh number. Results show that there is a range of wavenumbers and Ra with stable rolls. Subsequently, many more calculations for flows like those in Fig. 2.11 have been successful and many new flow patterns were found. Numerical computers not only assisted in the checking of experimental results, but they even have produced new results and resulted in many other studies over wide ranges of Pr.

Before discussing aspects of convection cells relevant to mantle convection, we mention three major results available on the web and not covered in the rest of this book. The first major result is a famous collection of studies involving the understanding of chaos. Simplified equations for a single convection cell in a box are

Fig. 2.11 Comparison of experimental results for cells initiated by inducing with a grid and their stability with theory. The closed curve is a prediction that a roll grows at right angles to the starting roll and the open curve is that a roll grows as a small angle to the starting roll. The experiments show + cross-roll instability circles stable rolls, x zig-zag instability, ‡ bimodal flow, and ≢ cross-roll instability inducing transient rolls leading to subsequent local processes (adapted from Fig. 6 in Busse and Whitehead 1971)

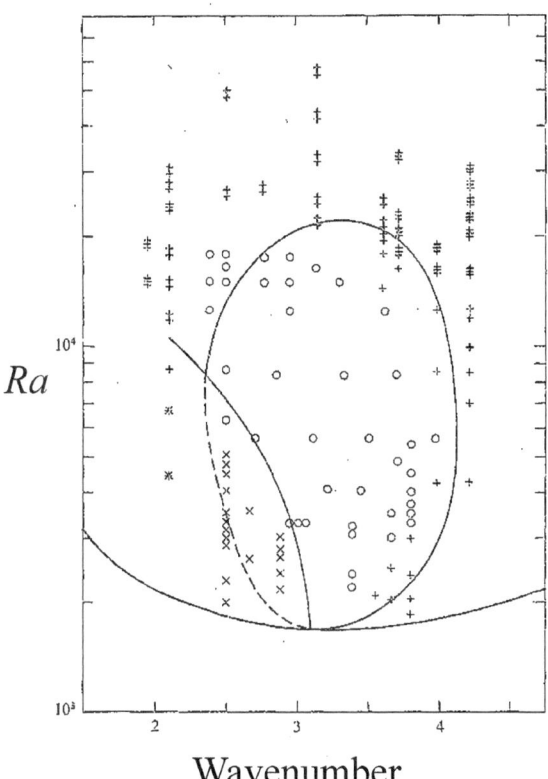

Ra

Wavenumber

called the a-b-c equations. They represent three variables: speed, temperature, and the temperature difference between the top and bottom. The meteorologist Ed Lorenz in the 1960s showed that equations for a-b-c can evolve to completely unpredictable results after a finite time. The new view of uncertainty possessed by these equations, called chaos theory, is cited as one of the most important fundamental contributions of the twentieth century.

A second major result was developed to estimate the maximum amount of heat flow (or in the case of pure turbulence, of kinetic energy production) that can exist in a wide range of possible cases. The approach is called "extremum theory" because it is one way of telling us what can limit the effects of turbulence for convection and other flows. The approach is mathematical in nature and quite technical, but the result is that we don't need to calculate fully developed, complicated turbulent flows to estimate an upper limit to the total heat or momentum they might carry.

A third major result consists of numerical models to determine convection at very large Ra, far above the modest range of Figs. 2.3, 2.4, 2.5, 2.6, 2.7 and 2.8. The methods are used also for weather and climate predictions and numerous other problems. The rest of this chapter tells how numerical work led to high resolution of boundary layer flows.

2.6   Large Ra Results and Boundary Layer Flows

In all these studies there is a flow of energy. In the top and bottom of convection cells there are boundary layers that allow the heat to flow upward between the cell and the surfaces. Let's clarify the heat flow in them. Below Rac, there is no motion, and the flow of energy upward is due to thermal conduction of heat upward between the bottom and top surfaces. The value of energy flow after convection cells exist is easily calculated.

The strength of the heat flow of motionless fluid is calculated by multiplying the thermal conductivity of the substance k times the temperature difference divided by fluid depth d. Therefore, the formula for heat flow per unit surface area is $k \times \Delta T/d$. This is valid below Rac but above Rac there is convection and heat flow is greater. We use a dimensionless value of heat flow called the Nusselt Number Nu. It is the value of heat flow divided by the strength by conduction. For Ra above Rac, the heat flow upward by convection cells becomes greater and the physical value of heat flow equals $Nu \times k \times \Delta T/d$.

The cells are driven by the flow of energy. Chapter 9 reviews the potential energy release of a mass in a field of gravity and its conversion to kinetic energy that does not take into account fluid flow. Here we start by considering the simple case of a mass that is forced to move up or down at a steady speed (for example, within a fluid) in a field of gravity. It requires work at a constant rate to keep the fluid moving. As time goes on, for a particle, you can calculate the value of work done over any time interval by taking the total work as in Chap. 9 and dividing it by the time interval. This expresses the rate of work that is being done and generally is called the power. This can "work" both ways. Power at a steady rate is required by a bike rider steadily pedaling up a hill and released when steadily coasting down the hill. Figure 2.12 shows the simple example that is developed in more detail in Chap. 9. Dense fluid enters a pipe at its top, flows downward in a field of gravity, and leaves at the bottom. A formula for energy flow from Chap. 9 is reviewed here.

Figure 2.12 shows denser fluid than surrounding fluid entering the top of a tube and descending a vertical distance where it leaves the bottom of the tube. The rate of fluid descending is a volume of fluid per unit time Q and the rate of potential energy decrease is $Er = g \times d \times \Delta\rho \times Q$. This liberated energy flow has the units of Watts (W).

For convection cells above Rac, the heat is carried by upward and downward by flow inside the rising and sinking portions of the cells (technically, we call this the "advection" of heat). A simple flow of a pair of cells is shown in Fig. 2.13 for $Ra = 10^5$. The convection cells are well formed. Notice that the temperature field has a sharp temperature change along the top and bottom boundaries. This region of large temperature change is called a "thermal boundary layer". The calculation of these flows as Ra gets larger and larger is well developed and heat flow is found to be proportional to the term $Ra^{\frac{1}{3}}$.

Although the patterns of convection cells are interesting, the energy flow within a convection cell gives a better view of how the cell is driven. Let's imagine that

Fig. 2.12 The power, or the rate of change of energy for fluid in a gravity field descending in a tube at a steady speed while immersed within a lighter fluid

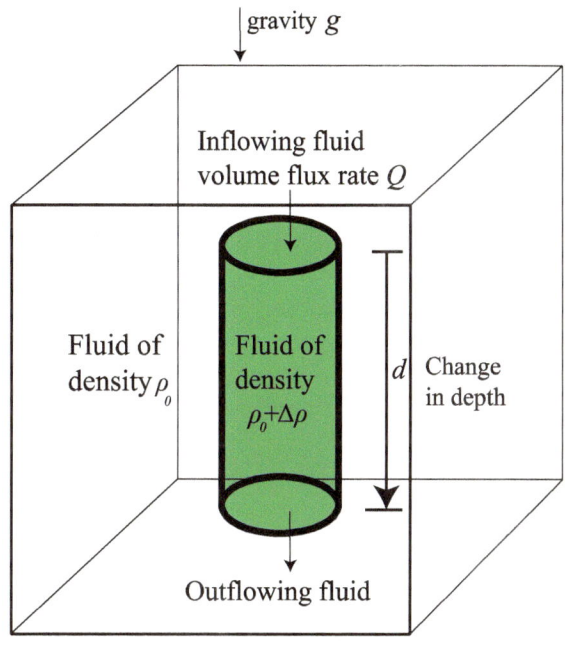

Rate of change in potential energy $= g \times \Delta\rho \times d \times Q$

the flow of energy is an invisible skeleton that organizes and underlies the entire process. The convection cell is propelled by lighter fluid with higher temperature at the bottom and heavier fluid with colder temperature along the top (Fig. 2.13). Energy in the form of heat flows in at the hot bottom by thermal conduction. The warmed fluid at the bottom is swept sideways, warming as it moves, and is focused into an ascending column of hot fluid (hotter than average, that is). This rising hot fluid brings up a lot of energy in the form of heat. At the top of the column, the hot fluid spreads out so that the isotherms look like the top of a mushroom. At the top, the surface is colder than any fluid below it and so the warm fluid in the mushroom is cooled as it spreads out laterally. This fluid then becomes cold enough to become part of the cold fluid that sinks along both sides. Along the top, energy in the form of heat flows up and out of the cell by thermal conduction in the top boundary layer. Another example of an energy-driven flow is explained in the tutorial in Chap. 9.

We are almost ready to move on to the giant engines of Earth. Remember that scaling allows us to extend results to large sizes. Before moving on to Chap. 3 which tells how we learned about the Earth and convection, here is one more calculation for the convection cell driven by uniform heat production within the fluid in Fig. 2.14. Heating like this within a real fluid might occur in several ways. One way uses internal heating by an electric current passing through an electrically conducting fluid. Another way has the fluid absorb radiation, as in our atmosphere and in my PhD thesis work. The third is that there is radioactive decay within the material, which

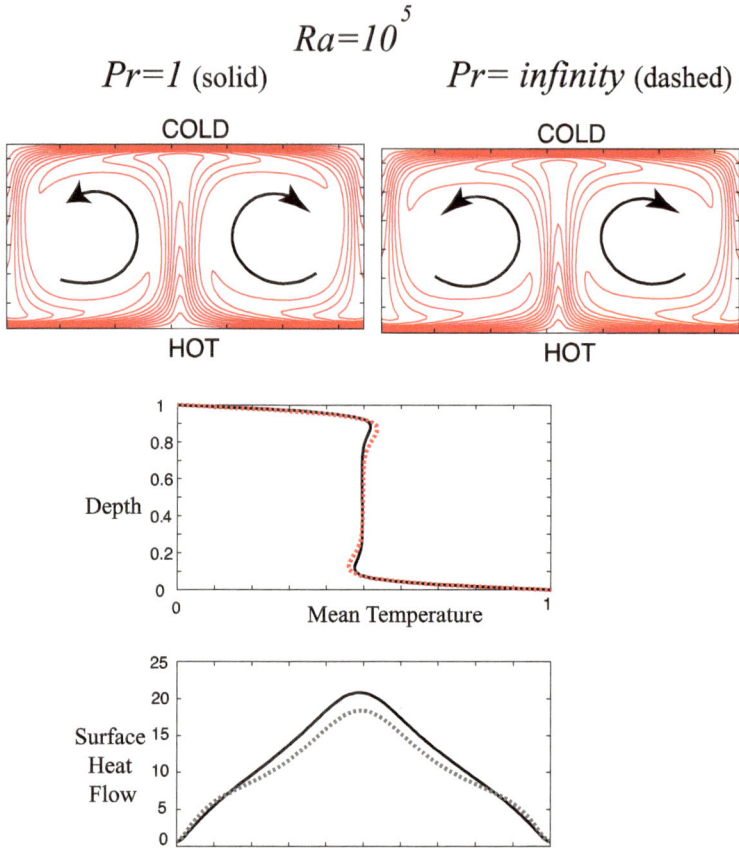

Fig. 2.13 Top panels: the circulation and isotherms (contours every 0.05) for temperature in a pair of convection cells at Ra = 10^5 for both Pr = 1 (solid) and Pr = ∞ (dashed). Middle panel: profiles of average temperature at each depth for both cases showing the top and bottom boundary layers. Bottom panel: the heat flow across the top of one pair of cells with the ascending flow in the mid-depth for the two values of Pr

probably occurs within Earth. The numerical calculations for Fig. 2.14 have the heat flow along the bottom and sides set to zero, so the internally produced heat must rise and escape by conductive cooling along the top. We see that just below the top is a boundary layer of cold fluid that collects and sinks because it is denser than the fluid below it. The descent occurs through three cold sinking columns of fluid that sink all the way to the bottom and spread out. The fluid then slowly rises and warms up because of the internal heat production.

We are not quite finished because the production of mechanical power in convection is especially interesting. The power, which is the rate of potential energy release, is found by calculating or measuring how quickly the lighter fluid flows upward and the heavier fluid flows downward over the entire length of the vertical motion, and

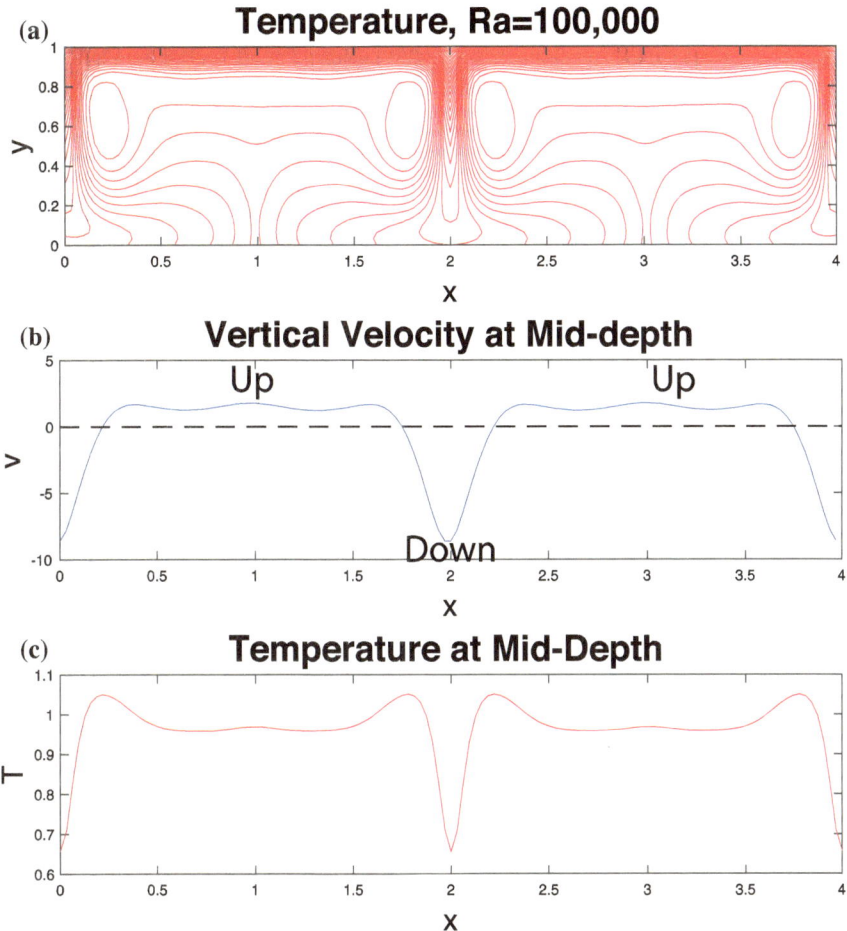

Fig. 2.14 Two-dimensional numerical calculation for convection with uniform internal heating and an insulating bottom boundary for $Ra = 10^5$, $Pr = \infty$, and a length to depth ratio = 4. **a** This vertical section shows 30 isotherms of the temperature field. **b** Vertical velocity at mid-depth. **c** Temperature at mid-depth

then multiplying it by the force of gravity and temperature change times expansion coefficient. Using the numerical program, one multiplies the speeds by the density difference at each point and then adds up this product times gravity acceleration over the depth over the entire cell. Using calculus, we call adding it all up "integrating" over the region. In general, it was realized early in the study of convection that for these convection cells, the rate of potential energy release is proportional to the vertical flow of heat. This is one reason that led Malkus to suggest that the convection naturally would form cells that maximize the vertical flow of heat, since they also maximize the rate of potential energy release. When I was a student, there had been several years of great excitement over the thought that this type of consideration could lead to great breakthroughs in the study of turbulence. The "holy grail" that maximization of heat flow or some other flow of a driving agent would lead to a general solution for theoretical turbulence was never found however, and we all moved on.

Here is a calculation of the flow of mechanical energy of a convection cell. It is like calculations in the books by Davies (2001) and another book by Turcotte and Schubert (2002). Let's estimate the energy flow for a simple convection cell. To apply this concept to a convection cell, we assume that one parcel steadily flows down from top to bottom and the other up from bottom to top. We assume that the center of the cell in Fig. 2.13 has a temperature difference from the average temperature of exactly $\Delta T/2$. The total power driving the vertical flow in two tubes (up and down) as in Fig. 2.12, is $\text{Er} = 2 \times g \times \Delta \rho \times d \times Q$. The power liberated by a convection cell with flows that produce a constant change of potential energy is the same $\times g \times \rho_0 \times \alpha \times \Delta T \times d \times Q$ and the units are joules per second, or Watts. The volume flow rate Q for a two-dimensional convection cell like the one in Fig. 2.13 is the speed u times the length the cell extends in the third direction (into the page) l_{ex} times the thickness of the boundary layer b. To make this the power per unit surface area of the cell, this product is divided by the surface area of the cell (the length of the cell L times l_{ex}). The power liberated per unit area of the chamber is therefore $u \times b \times g \times \rho_0 \times \alpha \times \Delta T \times d/L$. Now there are two things in this formula that don't have a value, the speed u and the thickness of the boundary layer b. We need formulas for them. To produce them, we incorporate another important component to the energy budget, and that is how the potential energy production rate is balanced. For these cells, we assume the vertical parcels are sheet-like rather than cylinders and the production is balanced by viscous dissipation. Without going into too many details, this is simply proportional to viscosity μ times speed squared times l_{ex} divided by fluid depth squared and per unit area of the cell it is $\mu \times u^2/L \times d$. Finally, the connection between speed and boundary layer thickness is approximated by the formula $b = (\kappa \times L/u)^{1/2}$. For a cell with $L = d$, the heat flow per unit area is given by the speed times the specific heat of the hot and cold flows times the thickness of the parcels b. The formula for heat flow reduces to $\frac{k*\Delta T}{d} \text{Ra}^{1/3}$ W. The overturning speed reduces to $u = \frac{\kappa}{d} \text{Ra}^{2/3}$.

This heat flow relation that is predicted by energy conservation has the Rayleigh number raised to the same exponent 1/3 mentioned a few paragraphs ago where it was also told that theoretical and numerical studies of convection cells generally have found a range for the heat flow exponent from 0.25 to 0.33 as Ra is larger and larger. Measurements were done from the 1930s onward and the mid-1950 experiments by Malkus and all have various values of the exponent. Many other theories are worked out in detail in various textbooks and websites, and we won't cover them here except to mention one very interesting calculation that is very simple. If we consider a layer of fluid at uniform temperature that has a top surface that instantly has a decrease in temperature, then the decrease in temperature will thermally conduct down into the material into a thin region with a thickness that increases over time. The inverted version of our milk experiment. The advance of the thickness is proportional to a well-known formula $\sqrt{\kappa \times t}$. If we assume this thickness advances until the

Rayleigh number defined by this thickness becomes large enough to reach a fixed value like 100 or 1000 or even Rac, then a formula for heat flow that is proportional to $Ra^{1/3}$ is the result. We can also find a similar relation if we assume the thickness advances until the linear growth rate is the same as the time that has already elapsed. Both calculations show how linear theory can be combined with some physical consideration to produce some estimates of heat flow and potential energy release rate. Generally, these calculations, although stimulating, are approximations rather than proven factual results.

We will leave the topic of cellular convection now because the energy and force balances presented here were also found to apply to the motion of the great plates on Earth. This will be shown in the next chapter.

2.7 Summary

We will move on to more practical considerations by looking at other examples of energy-flow structures that involve motion within the Earth. Before doing that, summarizing:

- The cells and many more complicated flow structures are driven by a flow of energy. For convection cells, the flow of thermal energy produces flow through buoyancy. This produces mechanical energy production, and this is dissipated by friction.
- We introduce dimensionless numbers, particularly the Rayleigh and Prandtl numbers. There are many others. Shear flow, for example, is governed by Reynolds number, which is probably the most famous dimensionless number. The story of its discovery and the development of understanding is long and more complicated than convection's story. Some, but not all dimensionless numbers relate to energy-flow structures.
- The numerical examples presented here are simple compared to many calculations that can be done today. Fully three-dimensional flows with quite turbulent flows have been successfully completed over wide ranges of the governing dimensionless numbers. Convection cells in spherical models of stars, our sun, our atmosphere, and planetary atmospheres have also been calculated.
- The flow of thermal energy in convection cells is generally proportional to $Ra^{1/3}$.
- Another body of studies (pursued by Malkus as we described, but also pursued by many others) asks whether there is one underlying principle governing turbulent eddies. Humanity has not arrived at any comprehensive discovery, although people still use this approach to find limiting features of some aspects of turbulence.

References

Bénard H (1901) Les tourbillons cellulaires dans une nappe liquide. -Méthodes optiques d'observation et d'enregistrement. Journal de Physique Théorique et Appliquée 10(1):254–266

Busse FH (1978) Non-linear properties of thermal convection. Rep Prog Phys 41(12):1929

Busse FH, Whitehead JA (1971) Instabilities of convection rolls in a high Prandtl number fluid. J Fluid Mech 47(2):305–320

Chen MM, Whitehead JA (1968) Evolution of two-dimensional periodic Rayleigh convection cells of arbitrary wave-numbers. J Fluid Mech 31(1):1–15

Davies GF (2001) Dynamic Earth: plates, plumes and mantle convection. Cambridge University Press

Getling AV, Brausch O (2003) Cellular flow patterns and their evolutionary scenarios in three-dimensional Rayleigh-Bénard convection. Phys Rev E 67(4):046313

Jeffreys H (1926) The stability of a layer of fluid heated below. Phil Mag 2:833–844

Malkus WVR, Veronis G (1958) Finite amplitude cellular convection. J Fluid Mech 4(3):225–260

Newell AC, Whitehead JA (1969) Finite bandwidth, finite amplitude convection. J Fluid Mech 38:279–303

Palm E (1960) On the tendency towards hexagonal cells in steady convection. J Fluid Mech 8(2):183–192. https://doi.org/10.1017/S0022112060000530

Rayleigh L (1916) On convective current in a horizontal layer of fluid then the higher temperature is on the under side. Phil Mag 32:529–546. Also, Sci Papers 6:432–446, Cambridge, England

Turcotte DL, Schubert G (2002) Geodynamics. Cambridge University Press

Whitehead JA (2003) Laboratory studies of mantle convection with continents and other GFD problems. In: Recent research developments in fluid dynamics. Transworld Research Network, Kerala, India

Chapter 3
Energetics of Mantle Convection

Abstract Mantle convection and the production of the great tectonic plates described in Chap. 1 have energy flow that is very similar to the laboratory and theoretical convection cells in Chap. 2. However, the overall structure of mantle circulation has dramatic similarities and distinct differences from the convection cells described in Chap. 2 that exist in a fluid dynamics laboratory. Despite all the differences, both convection cells and mantle convection are energy-flow structures driven by the flow of thermal energy (heat) upwards. This gives a brief history of how data from the ocean floors produced the concept of the energy flow that propels plate tectonics. The discovery of subduction and heat flow measurements led to a general acceptance of convection as a model that drives plate motion. The boundary layer concept is explained as a simple model driving the plates. The general flow of heat energy and its role in plate tectonics and mantle convection is quantified.

3.1 Heat Flow and Mantle Convection

What is the link between convection and mantle motion? First, it was already known that ocean floor depth increased on both sides of the undersea mountains that were found as the trans-Atlantic cable was laid by the Great Eastern in 1866. Ocean depth typically changes from less than 3 km at ridges up to 5 or 6 km deep in some places that are now interpreted as "old basins". The picture was put forth in the late 1960s that the plates are cooling and thickening as the newly formed plates move away from the ridges. That is, the newly cooled plate material becomes "older" while being swept away from the ridge axis. Seismology had already located a layer of very strong material below the crust called the lithosphere. It had been known for decades before plate tectonics that a reflection layer at the base of the lithosphere is a prominent feature. The present interpretation is that the reflection is the base of the cold plate with a transition in physical properties as temperature changes from cold to hot below the ocean floor. Therefore, the plates are literally the cold boundary layers of a mantle convection cell, and this cold region constitutes the lithosphere sketched in the layered Earth in Chap. 1 (Fig. 1.9).

J. A. Whitehead, *Energy Flow and Earth*, SpringerBriefs in Earth System Sciences,
https://doi.org/10.1007/978-3-031-62694-4_3

Despite the general skepticism about continental drift in 1960, the idea that the mantle of the Earth is overturning with a cellular convection nature was not new. It had previously been advanced many times in the studies of the Earth and was especially prominent in Europe in the 1920s in support of continental drift. Later, numerous objections based upon the perceived stiffness of the mantle in the era 1930–1965 led most geologists and geophysicists to reject the idea of convection cells within the mantle. The ocean floor was simply too strong to allow continents to plow through it. A persistent objection also came from calculations that the spinning solid Earth has a bulge that is not explained by centrifugal force alone. It had led to the idea that the deep mantle was very strong and immovable. By 1965 or so, with the plate concept becoming clear to many people, the old possibility of mantle convection was being strongly re-introduced. Modern books with detailed discussions are Davies (2001), Turcotte and Schubert (2002), and Schubert et al. (2001). To help, by 1970 the bulge was shown to be simply a remnant of the wander of the pole of rotation from mass redistribution from mantle motions.

When I was a graduate student studying cellular convection in an engineering department, the idea was already being introduced to many of us in fluid mechanics that the plates that formed at the ridges were like the cold boundary layer in Figs. 2.8 and 2.14. Therefore, in 1968 I enthusiastically joined the mantle convection community as a postdoctoral fellow at the Institute of Geophysics and Planetary Physics at UCLA with Willem Malkus. Work on convection cells that are described in Chap. 2 only increased my curiosity, and 3 years later, when I joined Woods Hole Oceanographic Institution (WHOI), I was eager to learn about measurements of heat flow coming up through the ocean floor. Measurements around the global oceans were starting to show that the heat flow was greatest near ridges, and that was consistent with the concept of the thermal boundary layer. I was fortunate to be invited to be on thesis committees for students supervised by Dick Von Herzen at WHOI, who was one of the pioneers of ocean heat flow measurements. In 1976, I even participated in a cruise to view how the ocean heat flow measurements were taken. This was the second leg of the discovery of deep-sea hydrothermal activity in the Galapagos Islands. Heat flow out of the ocean floor is vital to the picture of energy flow and its role in forming the flows, but before telling of what those ocean floor heat flow measurements meant for mantle convection, I want to describe the procedure for measuring upward heat flow in the ocean floor.

To measure heat flow, a long vertical probe with a great weight at the top is lowered by the ship's winch to a position just above the ocean floor (floor depth is located by sonar) and then forced into the mud-like sediment that is found almost everywhere. This sediment is scarce at ridge centers but is present near the biologically productive equatorial Eastern Pacific. The dramatic verbal command to the winch operator, "full speed down × meters" placed the pre-positioned probe tip down into the sediment. I can't recall how many "×" meters down were required before the winch stopped, just a few, I think. Then, the probe is left in the mud for half an hour or so, after which it returns to the surface. Temperature records are taken by thermistors along the length of the probe. They typically reveal a downward temperature increase inside the mud that is small but easily measured of up to 0.1 °C (Ocean water temperature changes over time at those depths are generally smaller than 0.1 °C). From the small

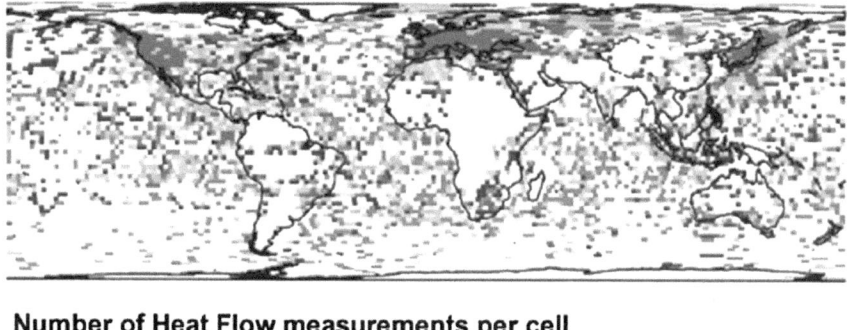

Number of Heat Flow measurements per cell

1   2   3 - 4   5 - 7   8 - 16   17 - 1095

Fig. 3.1 Density of heat flow measurements over the Earth. From Davies (2013) Fig. 5, used under Creative Commons CC-BY license

increase in temperature with depth in the sediment, the energy (heat) flow from inside the Earth is calculated by assuming that the thermal conduction of heat upward is steady. That is, the thermal conduction of heat upward from the hot mantle to the surface, even in the sediment, has reached a steady value over geological time. The value of the thermal conductivity of a sediment can be obtained by bringing a sample up to the ship and then measuring it later. Therefore, the heat flow is calculated by multiplying the value of the temperature gradient (the change per unit of depth) by the mud thermal conductivity. Well-taken results are convincing but sadly the global coverage of heat flow measurements is irregular (Fig. 3.1) because of the expense of the entire operation. It uses a few hours of expensive ship time. Therefore, ocean bottom heat flow measurements are dwarfed by the number on land. In fact, the total number of ocean floor heat flow measurements is still probably less than 2000.

Values of heat flow out of the floor of the ocean showed that heat flow near the ridges tends to be largest and there are smaller values away from the ridges. This is consistent with two moving plates that are produced at a ridge and becoming cooled as they spread apart. It soon was clear that if the heat flow is plotted as a function of the age of the plate instead of distance from the spreading center, then different plates had heat flow values that overlapped well. How did this idea come about? A very simple and striking model of heat flow that is well-known to thermal engineers is behind it.

Let's imagine a model that is too simple for plate tectonics but is needed to understand heat flow on the ocean floor. Consider a motionless material that starts with a constant temperature at all depths and then is suddenly given a colder temperature at a top boundary. The material becomes cooled by thermal conduction and the cold region becomes deeper with time. If the thermal conductivity is a fixed value and not dependent on temperature or pressure, the deepening as time progresses of the cold region is well understood. Without going through the details of the mathematics, a

beautiful formula shows that the penetration distance of the profile of low temperature near the top gets deeper and deeper as time goes on with the same shape. Even
better, this shape deepens at exactly the square root of time.

Now imagine a second case where the material is not motionless. Instead, hot
material is fed laterally at the left side at a constant speed and moves toward the right
(Fig. 3.2). The top surface is set at a cold temperature everywhere. In this model,
the left side corresponds to an ocean ridge. We desire to see how the hot mantle is
cooled by flowing away from the ridge. The cold region of the mantle extends by
thermal conduction down into the moving hot material. Since with steady lateral
movement, the distance along the plate is equal to the speed times the time that the
material has existed. The profile of low temperature advances downward exactly like
the advance with time for the stationary plate in the preceding paragraph, but now the
downward motion of coldness increases toward the right. That is, the mathematical
profile extends to deeper depths for the moving plate further and further away from
the left with the square root of the distance as sketched in the bottom panel of Fig. 3.2.
The inverse square root law not only applies to the temperature field, but the flow of
heat out of the top is proportional to the inverse of the square root of the distance,
too. The top panel in Fig. 3.2 is a sketch of this. The heat flow along the ocean floor
in sediments can be measured by marine geologists. To compare around the entire
globe, we note that different plates spreading apart at different speeds will have the
same vertical profile if the temperature is plotted as a function of the age of the plate
away from the ridge rather than the actual distance. This age is easy to calculate since
it is the distance away from the spreading center divided by the speed of the plate.

To make the final connection of this simple cooling model with Earth, we define the
tectonic plate as the cold region that deepens away from the side as the material moves
away. Although the situation is greatly simplified and many aspects are ignored, this
model of a cooling material moving steadily away from a ridge works extremely
well. In fact, when heat flow for different plates all over the world is plotted as a
function of plate age rather than distance, the result is a beautiful collapse of data.
It was only 10 years after Turcotte and Oxburgh's (1967) paper that Parsons and
Sclater (1977) produced their famous graph that we show in Fig. 3.3. It shows the

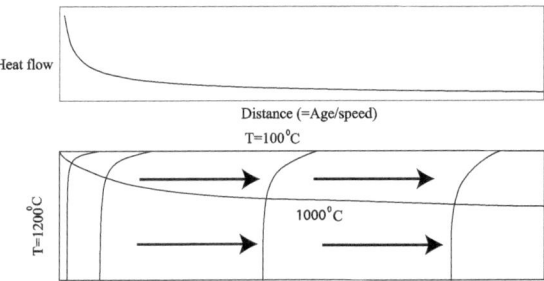

Fig. 3.2 Temperature change as a hot 1000 °C material moves steadily to the right with the top
set at 100 °C. Heat flow upward is greatest at the left and decreases to the right. The four cold
temperature profiles deepen with the square root of the distance from the left side

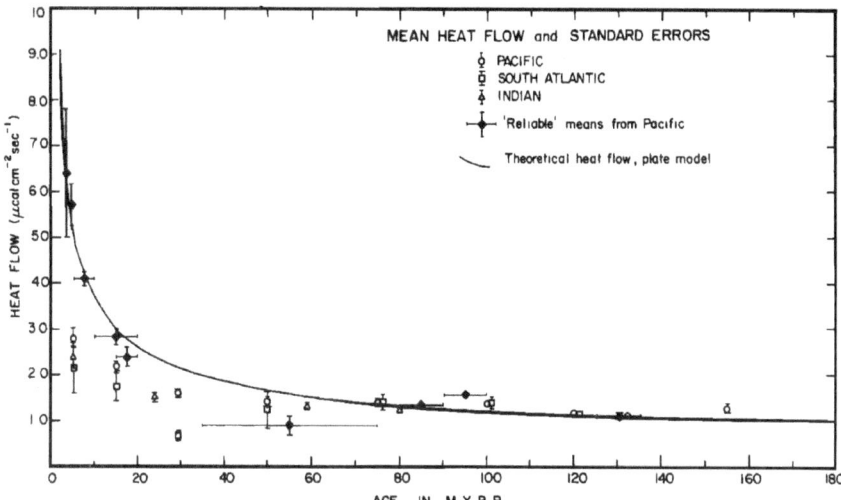

Fig. 3.3 Heat flow data showing how the flow decreases away from spreading centers. Parsons and Sclater (1977) Fig. 10, used under Creative Commons license

collapse for heat flow on plates in the Atlantic, Pacific, and Indian Oceans plotted against age (the spreading rates differed by factors of over 5). The simple heat flow along the top from a theoretical solution like that sketched in Fig. 3.2 is the solid curve that gives a good comparison. One of the authors, Barry Parsons, came to our laboratory and did convection experiments with me, and I was very excited when I found out how well Fig. 3.3 fits into the idea of mantle convection. Figure 3.3 also shows the upper portion of the energy flow out of the Earth that is associated with plate tectonics. This strongly illustrates that hot mantle material rises under ridges, spreads apart, and is cooled. The plates ultimately arrive at subduction zones and plunge into the mantle.

A modern map of heat flow out of the ocean floor worldwide is shown in Fig. 3.4. Because the data plotted as a function of plate age have proven to be so consistent, this map incorporates estimates (based on the inverse square root relation) for the flow in regions with few measurements. The overall picture has the largest values centered on the great ocean ridges and then decreasing with local plate age.

Next, since the plate moves away from the spreading center and cools, the lithosphere becomes denser from contraction from low temperature. We often say that the thickness of the cold, dense plate increases with age. Even though the plates are colder and denser than the mantle lying below, they float upon it. Therefore, the surface of a plate is shallowest at the spreading center, and it becomes deeper at older plate locations because the cold, dense plate is thicker. This change in elevation was already known before plate tectonic was understood, but when the ocean depth was carefully mapped out in different oceans in the mid-1970s, the fit of seafloor depth to the age relationship proved to be as impressively good as the heat flow fit (Fig. 3.5).

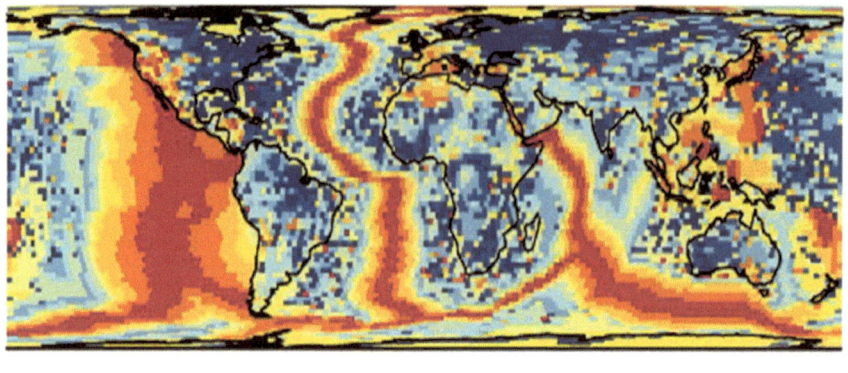

Final Estimate of Heat Flow (mW m^-2) (Area-weighted Mean)

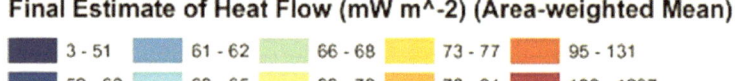

| | 3 - 51 | | 61 - 62 | | 66 - 68 | | 73 - 77 | | 95 - 131 |
| | 52 - 60 | | 63 - 65 | | 69 - 72 | | 78 - 94 | | 132 - 1237 |

Fig. 3.4 This map of estimated heat flow uses all the available heat flow data plus some geological interpretation. From Davies (2013) Fig. 7, used under Creative Commons CC-BY license

In fact, when one compares Figs. 3.3 and 3.5, it is obvious that the scatter of heat flow in Fig. 3.3 is greater than the scatter of ocean floor depth in Fig. 3.5. This means that (ironically) the variation of depth of the ocean floor (bathymetry, which is relatively easy to measure) is possibly a better indicator of energy coming out of Earth than direct heat flow measurements, which involve more difficult and expensive measurements. However, without the heat flow measurements as calibration, this model of cooling plates that spread apart would be speculative.

To summarize, the scaling relationships explained in Chap. 2 along with our understanding that hot material cools with time, indicates that the depth of the floating lithosphere surface changes away from the spreading center as the square root of age. The relationship between heat flow, bathymetric depth, and age proved to be so good that the relationship is now regarded as fact.

Getting back to my participation in this story, since this type of heat flow measurement cannot be taken in an ocean location where a bottom sediment is absent, coverage of heat flow data is rare near ridge centers. I participated in a study in the Galapagos where I learned how to measure heat flow in the ocean floor. The location was important because extremely large biological productivity has produced a thick sediment cover even over the spreading centers. Our project showed very large values of heat flow as we expected, but excitement over the discovery of new life forms gathered most of the attention.

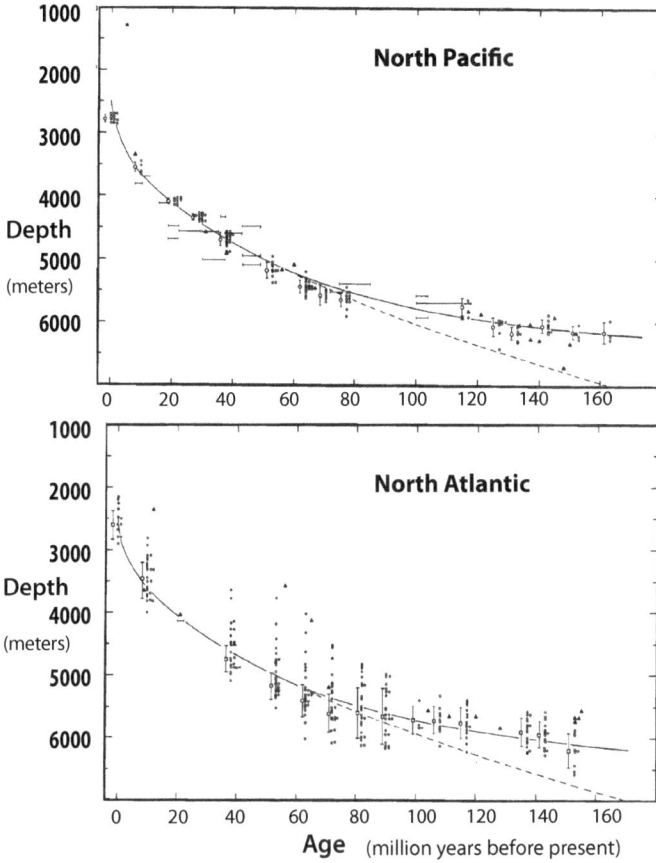

Fig. 3.5 Ocean floor depth as a function of plate age for the Atlantic and Pacific Oceans. From Parsons and Sclater, (1977) Fig. 15, used under Creative Commons CC-BY license

3.2 Cellular Convection and Plate Motions

Next, let's look more closely into the mechanics of the flow. The plates flowing away from spreading centers can be thought of as the flow of plates over a viscous fluid. The plates must have a force of drag along their bottoms, and the sum of the drag forces can be estimated and compared with the total forces of buoyancy from the sinking plates at subduction zones. A calculation of drag needs to use a value of the viscosity of the mantle below the plates. This value is unknown in general, but an estimate exists from other geological observations. The land surface has been rising in locations where after the great glaciers melted away at the end of the last ice age. This general area of study, called Fennoscandia uplift, can produce estimates of mantle viscosity (cited even in Wegener's book). The studies assume that various values of viscosity and density are distributed in different layers below the surface. The resulting values of viscosity were found to allow decent agreement between plate drag and buoyancy driving. Unfortunately, any close comparison in detail between such simple calculations and mantle convection has been challenging because the uplift regions are mostly situated near the continents that had the glaciers. They might not represent the mantle throughout the Earth.

Although the connection between lithospheric plates and convection is now widely described by review articles, books, and on the web, at first the connection between plate tectonics and convection was slow to become accepted. In the mid-1970s as all the information became available, there was a lot of discussion about whether plate motions and deeper mantle convection were connected to deep convection cells. Slowly, the consensus emerged that the concept of cellular convection as a model for mantle convection has strengths and weaknesses.

We have described some of the strengths already: 1. Plates are the cold boundary layer spreading apart from the middle. 2. Subduction is like the sinking boundary layer. 3. The speeds (discussed further on in this chapter) are close to what theory would indicate but only when using certain values of mantle viscosity.

Unfortunately, the weaknesses are still numerous. 1. The flow patterns in the horizontal direction of any plate differ from the cellular convection patterns such as rolls, polygons, or spokes. 2. The plate motion is unlike the fluid surface of a convection cell since laboratory fluids do not move as a rigid plate (of course) (In Figs. 2.8 and 2.14, for example, the surface flow of convection cells along the top is not like a rigid plate at all). 3. In the vertical direction, the cold sinking boundary layers of cellular convection move straight down vertically and are not like the tilted subducting slabs of plates (Fig. 1.6). 4. The plate material at both sides of the spreading centers moves apart with almost perfect symmetry and this is not necessarily true for the spoke pattern in convection. As far as I know the spokes often slowly change. 5. There are lots of reasons to think that the mantle viscosity varies greatly over different regions. The base of the plates might be a low-viscosity layer, the deep mantle might have very high viscosity, and the dynamics of the margins of the rigid plates cannot be viscous at all. 6. Laboratory experiments with fluid having large viscosity variation produced a sluggish layer of very cold viscous fluid along the top that has very small motion. My present view is that the plates themselves must possess some characteristics of both a solid and liquid material to reproduce plate tectonics in detail.

The numerous features of plate tectonics have not hampered the efforts to produce more than thirty numerical models for mantle convection/plate tectonics. These models are calculated using large computers that numerically step forward the energy and flow equations. They generally incorporate many aspects of the structure inside the Earth. They are impressively good, but they differ from each other in the expected speeds and either temperature distribution (if heat producing within is imposed), or heat flow (if temperature of the lower mantle or core is imposed) with ranges varying by many tens of percent because of five principal things. First, the temperature distribution and flows in the mantle must be assumed at the time of start of the calculation; second, there is always the absence of one clearly agreed distribution of viscosity over depth; third, the three-dimensional nature of plate motion is not always included; fourth, contributions from plate elasticity are included in some and not in others; fifth, influence of continents is included in some models and not others; sixth, the contributions from very deep flow below the phase change between upper and lower mantle are included in some and not in others; seventh, generally the distribution of temperature in the lower mantle is unknown and unlikely to adjust as quickly as needed for the model; and eighth, possible upwelling from the deep mantle is included in some of them but the actual upwelling is not known. Some baffling aspects remain. The 3-d pattern of the return flow in the mantle below the plates is virtually unknown at present, a calculated return flow is not verifiable and certainly it is not mapped. There is still uncertainly of the distribution of viscosity in the mantle. In addition, although seismology gives a picture of the density and fabric of the crystals within the mantle, the methods to measure the actual motion are all poorly developed so far. Although viscosity distribution remains uncertain in detail, numerical models are becoming increasingly powerful, and some resolution might occur soon. The deformation and stresses are all still being studied, along plate edges, especially where the solid-like plates bend before plunging into subduction zones. The shallow and deep focus earthquakes that indicated in the 1960s that the plates sink deep into the mantle still remind us that the local flow around slabs is not simple viscous flow. Many of the problems exist because we are uncertain about stress-flow laws. We would like to learn more about them. Others exist because of a lack of data. Because of all these factors, the evolution of structural plate evolution over 2 or 3 billion years of Earth's history cannot be calculated now.

However, the view that some of the features of cellular convection are important factors in plate motion has persisted. Principally, many would agree that mantle convection is driven by the same energetics as convection. The cold slabs sink and drag the plates along. Rising at spreading centers also propels the plates but to a smaller extent. The plates are like cold boundary layers and develop in a similar way to those boundary layers in Chap. 2. Therefore, the energy flow of the convection cells in Chap. 2 and the mantle motion in this chapter are similar. A primary objective of this book is to remind everyone about this.

In summary, convection cells including the flow of energy are useful concepts as a model of mantle convection and plate motions even though a detailed study of mantle convection is loaded with uncertainties. I hope that this book so far has explained why we use the Rayleigh number as a criterion for the vigor of mantle convection. Let's estimate a possible value.

Recall the formula $\mathrm{Ra} = \frac{g \times \alpha \times \Delta T \times d^3}{\kappa \times \nu}$. To estimate a value for the mantle let us start with the simplest case and assume there is no internal heat production. Four physical properties (gravity, thermal diffusivity, the temperature variation, and the coefficient of thermal expansion) have reasonably well-known values. Two other variables, the viscosity and the depth of the convection region are poorly known. As mentioned earlier, the best measurements of mantle viscosity use data from the post-glacial rebound (the gradual rise of Earth's surface after the glaciers melted at the end of the previous ice age). Those studies used a model that is a vertical profile of viscosity with depth in the mantle with numerous layers. The most highly developed models also include profiles of density and elasticity that are as realistic as possible. Generally, they use a fluid flow model that slowly deforms, and the results are compared with data sets. Even the best models don't have the resolution to produce any clearly confirmed viscosity profile with depth, but results seem to be within the correct order of magnitude to support the idea that the mantle can be regarded as a viscous material with layers of higher viscosity at greater depth, a thinner layer of lower viscosity at depths of 100–200 km and then higher viscosity and brittle-elastic features near the top.

Using the ranges of viscosities and depths of mantle convection, the estimates of Ra over the past 40 years range from 10^6 to 10^9. This range is large and likely to remain so until new data help us. The depth of penetration deep in the mantle is uncertain for subducting slabs, that is, for the vertical overturning of mantle convection. Whole mantle convection models and layered convection models both gathered strong supporters in the scientific community over my scientific lifetime (so far). The existence of phase changes in the layered mantle contributes some uncertainty to this issue. The changes in composition of the plate near ridges produce a variety of different models. We are not sure how deep the return flow is, and even what exactly constitutes the mean flow of the mantle (and even how to produce a convincing measure of it). The deep flow return flow from subduction to spreading centers is probably complicated in all three dimensions, but no data set exists that can produce robust vertical profiles or maps of the return flow. In addition, the exact manner of weakening at the margins (so that the plates remain rigid) is still uncertain.

The possibility of smaller scale convection cells below the plates has gathered interest. Frank Richter and Barry Parsons constructed experiments with my help in our laboratory in the early 1970s to explore whether small convection cells could lie below the oldest plate locations, which seemed to have higher heat flow than expected (Richter and Parsons 1975). They could find some slim supporting evidence. A recent study by Lees et al. (2020) gives a new perspective to this issue. In their paper, I immediately recognized (here, Fig. 3.6) the spoke flow patterns from our early 1970 laboratory experiments (Fig. 2.10). They utilized modern computer simulations of convection under the African plate. Their model used a layer of fluid under the plates with constant viscosity with heat supplied uniformly along the bottom to simulate the flow of heat from deeper in the Earth into the bottom of the layer. The study is a good example of the overlap between fluid mechanics and mantle convection. The results are compared with detailed maps of Northern Africa with the suggestion that some of the volcanism there is correlated with the structures in Fig. 3.6.

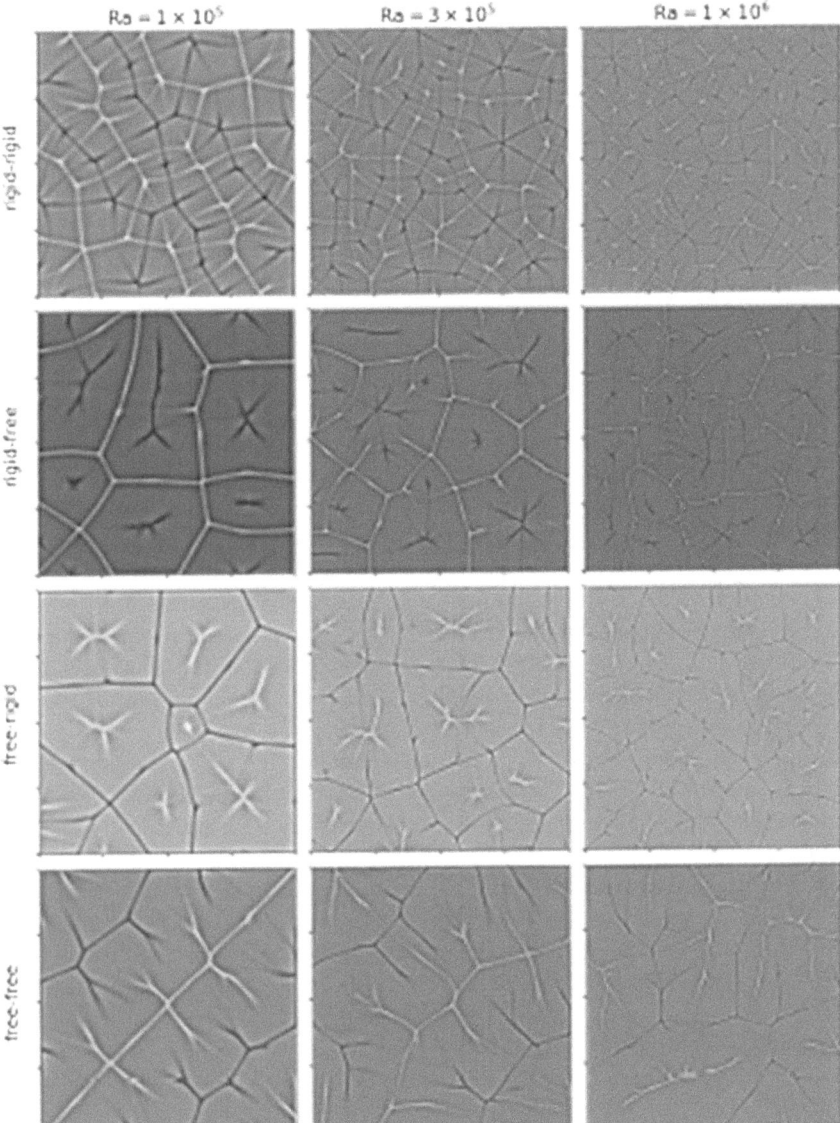

Fig. 3.6 Numerical calculations of convection under the African plate. These are "artificial shadowgraphs" of the 12 numerical experiments of thermal convection in a rectangular box. Each column is at a given Rayleigh number, and each row is at a given choice of whether the fluid at the top and bottom surfaces adheres to or slides along them. In each case the first word corresponds to the top boundary, and the second word to the bottom boundary. For example, free-rigid refers to a free-slip top and rigid (fluid adhering) to the bottom boundary condition. In Fig. 1, Lees et al. (2020) used under Creative Commons CC-BY license

3.3 Energy Flow

Let's pursue the topic of detailed comparison of models and data to focus on the path of energy flow. Although plates and subducted slabs have some structural features that are different from convection cells, the plates, and their subduction have sponta-neously produced the forces to drive themselves just like convection cells. Therefore, despite the hazy and complicated three-dimensional return flow in the mantle, its energy flow, which is emphasized in this book, is like simple cellular convection. Chapter 2 shows how the upward flow of thermal energy in convection cells produces a buoyancy-driven gravitational instability that produces the forces that drive the cell. The instability leads to a self-perpetuating flow. Furthermore, the sinking and/or rising thermal columns for both convection cells and the mantle provide the power through the steady release of buoyancy-driven potential energy. Let's perform a very simple calculation for mantle mechanical energy conversion to see this. First, let's calculate the rate of change in the potential energy of all the sinking slabs in the world. If a parcel of cold lithosphere were to instantly sink from the top to about 1000 km, then the calculation in the previous chapter for the sinking material is appropriate. Now this is applied to the mantle rather than a convection cell.

The change of potential energy for a unit-sized parcel that starts at the top and sinks a distance d to the bottom without a change in temperature from the example in Chap. 1 is $g \times \rho_0 \times \alpha \times \Delta T \times d$. The units are joules per cubic meter. As in Chap. 2, the rate of the liberation of this energy in Watts for all the subduction zones on Earth is equal to this change multiplied by the total volume flow rate (cubic meters per second) that feeds into subduction zones on Earth. To estimate this, we use a plate speed u times the total of all subduction zone lengths l times the thickness of a typical plate p. Therefore, the formula for energy flow (power) from potential energy liberation is $u \times l \times p \times g \times \rho_0 \times \alpha \times \Delta T \times d$. Using the following extremely approximate average values from observations of plate motions: a speed of $u = 10^{-9}$ m/s, length of subduction zones $l = 5.5 \times 10^7$ m, (approximately 55,000 km), a plate thickness $p = 10^5$ m (100 km), the gravity acceleration $g = 10$ m/s^2, an average density $\rho_0 = 4700$ kg/m^3, a coefficient of thermal expansion $\alpha = 10^{-5}$, a temperature difference between mantle and the surface $\Delta T = 1000$ °C and the depth of convection 1000 km or $d = 10^6$ m. The power is $26{,}000 \times 10^{-9+7+5+1-5+3+6} = 2.6 \times 10^{12}$ W. This value of mechanical energy production rate must be equal to power dissipation by viscous drag or other dissipative processes. The value of viscous drag has probably the widest range of uncertainty because both the viscosity distribution and the depths of the shear and the return flow regions are poorly constrained by geophysical data over the entire globe. Therefore, we use a simple model. For a plate of area A that moves over a layer of uniform viscosity fluid, the formula for viscous dissipation is $\mu \times A \times u^2/d$. Using the values for u and d along with the surface area of Earth of about 510×10^6 km^2 (5.1×10^{14} m^2), but leaving viscosity unspecified for the moment, produces a value of dissipation from drag of a plate that is $5.1 \times 10^{\,14-18-6} \times \mu = 5.1 \times 10^{-10} \times \mu$ with the units of Watts. Since this very simple shear flow is set equal to the rate of energy production, this produces an estimate of a value of μ of 5×10^{21} Pa s.

This example gives us a good picture of a crude order of magnitude calculation. Numbers can be picked that seem reasonable, and this one produces a value for viscosity equal to 5×10^{21} Pa s. How does this compare with other estimates? The magnitude lies between estimates of the value of mantle viscosity from direct measurement that range from a high of 3×10^{22} Pa s based on inferences from sinking speed of subducted lithosphere (Čížková et al. 2012), to a low that can be up to 100 times smaller, depending on assumptions about internal layering in the mantle using glacial rebound theory. Clearly these estimates are very sensitive to the depths of the return flow and shear zone under the plates, and neither is known from direct observation. The value $d = 1000$ km is used here, but how good is it? Perhaps the vertical distance of subduction and shear regions occupy much smaller depths than used here. The literature also has studies that suggest that the bases of continents supply a large amount of the drag that is not considered here, and finally perhaps the energy to deform the bending plates should not be neglected. Most large-scale numerical models of mantle convection (summarized for example by Bobrov et al. 2022) do of course recover a much better energy balance than this estimate. The models that produce velocities close to those observed for some plates generally possess low viscosity with large slips in a confined region just below the plates.

The previous chapter showed how the upward flow of heat is a fundamental part of the convection engine that drives cellular convection. Let's focus now on the sensible heat arriving at the surface of the Earth from below rather than mechanical energy. How big is this upward flow of heat? The flow of heat out of the ocean floor is readily calculated using the average ocean floor heat flow in Fig. 3.4. The worldwide average is 90×10^{-3} W/m^2. The ocean floor covers 2/3 of the surface of the Earth with a total area of about 510×10^6 km^2 (5.1×10^{14} m^2) so the total heat flow out of the oceanic floor is roughly 3×10^{13} W.

We need to add to this heat flow total the additional contribution of the upward heat flow up to the surface within the continents and from volcanos. The estimates of both are much less certain for various reasons. For one, the continents are thick, and the base of the continents extends deeper than the lithosphere so that the base seems to be colder than the mantle at those depths. Groundwater flow makes simple conductive estimates like those in the ocean floor less useful. The value of heat conducted upward through the base of the continents cannot be easily estimated. Some people even have stated that supplying heat to the base of the continents is unlikely. To complicate things, the distribution of radioactive trace material in continents is not measured thoroughly at numerous depths, so the distribution of radioactive heating within continents is uncertain. My present impression is that the community thinks that it is not negligible.

The overall flow of heat that is delivered to the surface of the Earth from volcanos is also hard to estimate completely. Obviously, each volcano concentrates the flow of heat that is conveyed up to the surface. Even though the volume of the volcano above ground level is known (if good maps exist, that is) the total rate of heat flow rate also depends on the total eruption time for each volcano. Such times are only crudely measured. A good eruption history relies on detailed geological maps and isotopic dating of samples. Even the history of surface eruptions is not enough, since some magma never reaches the surface. To make uncertainly even greater, flow rate estimates also depend not only on the value of magma flow but also on localized heat flow in the rocks near the magma-rich locations. These might involve substantial local heat flow loss to surrounding rock. We discuss more about volcanos in the next chapter. For purposes of estimating the global values of continent heat flow, we set the value at half the heat flow delivered by conduction through the ocean floor, or half of that we estimate to be 3×10^{13} W. Therefore, the grand total estimate for the entire heat flow out of the Earth of 4.5×10^{13} W. This value is about 20 times greater than the potential energy rate of release of the mechanical driving that we estimated two paragraphs ago (2.6×10^{12} W).

It is important to mention in passing, that the ratio of heat flow to the power of mechanical driving in laboratory convection is 5% or less. This percentage is likely to be much smaller than the ratio for the mantle. The mantle is much more mechanically efficient than the laboratory in this regard, because there are larger thermodynamic effects from great changes in pressure within the mantle. Pressure effects are important for mechanical energy conversion. For example, the liberated potential energy released by the plates is dissipated by frictionally produced heat, but including this heat production in a global budget must include some effects of pressure changes as the material moves up and down. The effects add and subtract from the internal energy of the material by compression and decompression. Therefore, these internal rearrangements of energy do not change the overall budget. No pressure effects are described in this book. In addition, melting and solidification might affect the balances. One simple message to take away is that no extra energy (heat) is produced by friction. These balances are important for convection in any compressible flow with large changes in density from top to bottom due to pressure. Although a good example is the mantle, a better example is a star, where compressional aspects are central.

Therefore, our picture of mantle convection is one that is driven by our well-known but invisible energy flow even though the exact mechanics is complex. Probably in early Earth, radioactive heating and cooling of the mantle led to a gravitational instability of the cold boundary layer (the lithosphere) at the top of the mantle that caused the plates to become the top boundary layers of mantle convection. Numerical models of mantle convection of the entire Earth over its lifetime are not possible even though present models can be made to have "realistic" features. The history of mantle convection is also challenged. Observationally, the development of the first mantle convection cells when the Earth was young might never be documented in detail. Data sets for studying mantle convection are helpful, but some questions are difficult to answer because all the present ocean floor is less than 300 million years old. Therefore, virtually all the older information about Earth's history lies only within the continents. They show the well-known complicated history of continent collisions with plate margins, continent collisions with other continents, and continent collisions with smaller chunks of continent material (as in the Himalayas). The continents also overtake islands and hotspots and become modified by volcanism.

Geochemists help to provide us with some views of the chemical changes of the mantle during such collisions by analyzing continental and ocean volcanic material. In all the complex history of continents and ocean floors, it is safe to say that clear constraints about large features in Earth evolution are hard to find. For example, the iron core formation is a puzzle. It all depends on when iron-rich meteorites arrived to form our Earth compared to iron-poor ones. Many scientists presently think that the core formed early in Earth's history over a 100-million-year period, followed or perhaps along with a melted mantle and collision/moon formation. An exact time of the beginning of convection is still unknown. We describe more about mantle convection and continents in Chap. 5.

Computer studies of mantle convection exist (Bobrov et al. 2022). One of the challenges is the uncertainty about the mechanical behavior of the plates themselves and about the distribution of viscosity within the mantle. In addition, if one wants to duplicate the real Earth with the model, the initial conditions for any calculation hundreds of millions of years ago are uncertain (The continents are known best. The plates have been reconstructed to some extent, but the mantle interior has not). Some of the most advanced computer models advance in time for hundreds of millions of years or possibly over a billion years, but since the Earth is over 4 billion years old, the early evolution of Earth is poorly modeled. Many numerical models of mantle convection are complex and include the effects of the internal layers and their associated phase changes. Incorporating viscosity variations seems to help, generally with a layer of very low viscosity under the plates and then with layers with much greater viscosity going down to the lower mantle. Many of the computer models are still two-dimensional because plate-edge dynamics is less certain (although great progress has been made for this challenge).

The overall energy flow is simpler to produce. A crude sketch of energy flow for either convection cells or mantle convection is shown in Fig. 3.7. This crude picture focuses on the flow of hot mantle rising under the plates and transition on the mantle material to lateral movement along the top. The mechanical energy driving the motion, shown by two dashed curves, is seen in laboratory convection cells as sinking locations rather than rising from a hot lower boundary. The sinking of the mantle by plate subduction is also like the dashed curves. Ironically, little direct measurements of the idealized deep circulation and return flow for the mantle exist to compare with the predictions. Seismology, along with measurements of the magnetic and gravity fields, have helped to identify patches of different propagation speeds that might be linked to compositional and thermal patches which were laid down by ancient subduction in the deep mantle. The three-dimensional flow pattern within the mantle needed to sustain the internal heat and energy flows is only roughly estimated from the data with little independent support from numerical models. In contrast, plate data gives us information that is quite precise about plate speeds and plate structures along the surface as in Fig. 1.8. Therefore, numerical projects that calculate details of plate evolution for hundreds of millions of years are still rare.

The depth where the heat originates within Earth is estimated using different models of motion of the upper mantle. The deep mantle and Earth's cores are also involved. We know something from geochemical constraints that are not covered in this book that give some clues about the location of radioactive material and theories

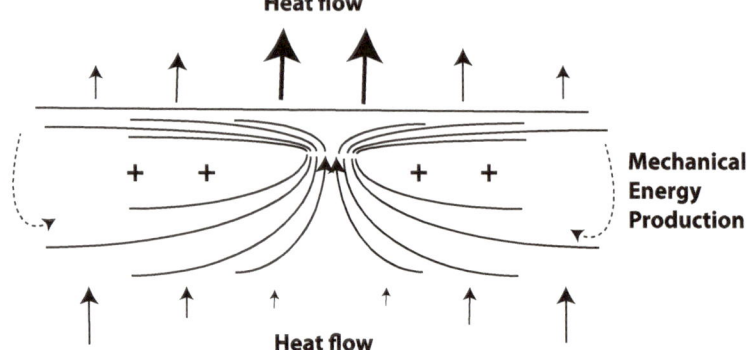

Fig. 3.7 The flow of energy for convection cells and the mantle

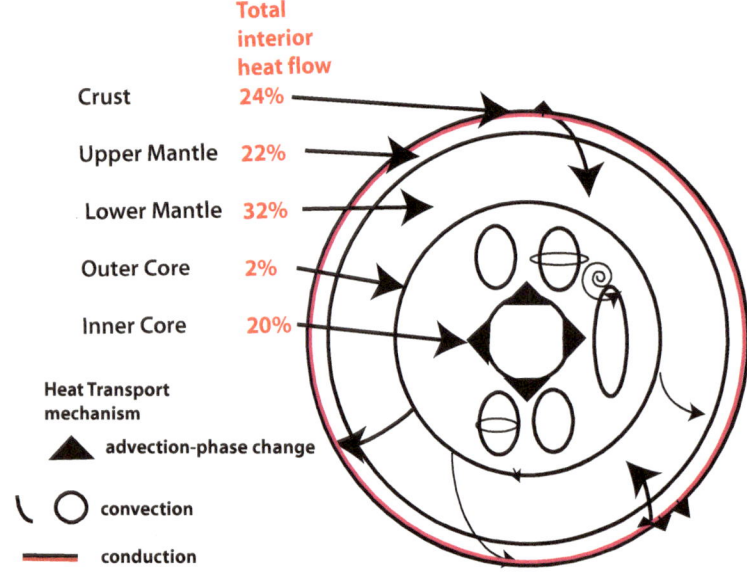

Fig. 3.8 Heat flow supplied by layers in the Earth (figure from Creative Commons)

of Earth's early accretion. Figure 3.8 is one estimate of the total values supplied by the different layers with sketches of the vertical transport of heat. Aspects of the energy flow and convection within the deeper layers are covered in some of the following chapters. We look next at regions of more focused upwelling and upward heat flow for Earth.

References

Bobrov A, Baranov A, Tenzer R (2022) Evolution of stress fields during the supercontinent cycle. Geodesy Geodyn 13(4):363–375. https://doi.org/10.1016/j.geog.2022.01.004

Čížková H, van den Berg AP, Spakman W, Matyska C (2012) The viscosity of Earth's lower mantle inferred from sinking speed of subducted lithosphere. Phys Earth Planet Inter 200–201:56–62. https://doi.org/10.1016/j.pepi.2012.02.010

Davies GF (2001) Dynamic Earth: plates, plumes and mantle convection. Cambridge University Press

Davies JH (2013) Geochem Geophys Geosyst 14(10):4608–4622. https://doi.org/10.1002/ggge.20271

Lees ME, Rudge JF, McKenzie D (2020) Gravity, topography, and melt generation rates from simple 3-D models of mantle convection. Geochem Geophys Geosyst 21(4):e2019GC008809

Parsons B, Sclater JG (1977) An analysis of the variation of ocean floor bathymetry and heat flow with age. J Geophys Res 82(5):803–827

Richter FM, Parsons B (1975) On the interaction of two scales of convection in the mantle. J Geophys Res 80(17):2529–2541

Schubert G, Turcotte DL, Olson P (2001) Mantle convection in the Earth and planets. Cambridge University Press

Turcotte DL, Oxburgh ER (1967) Finite amplitude convective cells and continental drift. J Fluid Mech 28(1):29–42

Turcotte DL, Schubert G (2002) Geodynamics, 2nd edn. Cambridge University Press

Chapter 4
Energy Flow and Magma Generation

Abstract Energy in the Earth can rise from deeper levels in the mantle where either excess temperature or compositional variations cause ascent to the surface. These upwellings are generally circular and localized and are often associated with magma generation and volcanism. The mechanics of localization is reviewed for either lower or higher viscosity rising fluid. The energy budgets of spreading centers, hotspots, and island arc volcanos are presented.

4.1 Regions of Magma Production on Earth

So far in this book, mantle convection is described in terms of a single fluid. The picture of large convection cells depends on a mantle being composed of crystals that slowly deform with a bulk viscosity. The deformation is generally assumed to be viscous in character (like flowing honey) and it allows the lithospheric plates to move over a slowly deforming mantle. These large-scale convection patterns move material around inside the Earth, but the convection itself doesn't involve the chemical differentiation of materials within the Earth. We all know, however, that the separation of different materials to produce rocks, soils, water, and salt in the ocean and nitrogen, carbon dioxide, and oxygen in the atmosphere are all important over the history of Earth. At the beginning of Earth's history, we can imagine that the early Earth had layers of accumulated material from repeated meteorite bombardment and possibly large amounts of melted liquid, including iron that sank to form the core. The most effective mechanism for separating different chemical components for early Earth occurred by cooling a liquid magma ocean so crystals accumulated. At some stage, the mantle formed and was composed of crystals that lay above a liquid core comprised of denser iron with other metals dissolved in it. After the mantle had formed, the surface crust accumulated by the flow of a melted liquid that presumably migrated upwards through the matrix of mantle crystals and erupted in the form of volcanism. What drives that volcanism? The story of these processes, and how the fluid mechanics was beginning to be understood, happened as I watched and participated (a bit). The general picture is that with mantle convection present, the mantle material under

J. A. Whitehead, *Energy Flow and Earth*, SpringerBriefs in Earth System Sciences,
https://doi.org/10.1007/978-3-031-62694-4_4

ridges with large pressure at great depth becomes swept upward and experiences lower pressure as it rises. Since the mantle material is an aggregate of various mineral crystals, there would be some chemical components that would melt out of the crystals because in general, most materials melt at lower pressure and solidify at higher pressure. The most common places for melt formation are at places where the rising of the mantle material produces a lowered pressure. The melted liquid would naturally form around the crystals and either would collect in small pockets of liquid or spread out around the crystals in a connected network. One or the other depends on the surface energy of the material. Liquid in small pockets would remain fixed, but liquid in a network would flow up or down next to the crystals by porous flow, because the melted liquid and crystals generally have different densities. Therefore, the porous flow of the liquid in a crystal matrix is quite common inside the Earth (and other planets, too) given sufficiently high temperature. It is really that simple. This porous flow allows different chemical materials to separate out from the mantle. All that is required is that mantle material can rise to a sufficiently low pressure at a warm enough temperature to have something melt off. There is seismic evidence of the liquid-forming region existing below spreading centers Toomey et al. (1990). Any place on Earth where such liquid melt has arrived at the surface from deep in the Earth is manifested by volcanoes. Not every melted liquid does this. Some compositions might be dense enough to sink down instead of up. In any case, volcanos are great places to study the results of such flows. They end up providing many of the incredible chemical and compositional variations of Earth. They even provide clues to the original formation of the continents and oceans.

Geologists collect many examples of volcanic activity on the continents, but on the ocean floor samples are much rarer. This difference is important. Therefore, it might be surprising (at least to the public) to know that the greatest volume flow rate of magma is not at any known volcanos, but at mid-ocean spreading centers. More surprising is that spreading centers don't have significant volcanic elevations. In fact, the production of ocean crust was not understood before the 1960s. When cameras were encased in pressure vessels with strong windows, they photographed lava forms along the Mid-Atlantic Ridge that are found in surface volcanos like Hawaii. It was soon realized that magma solidified and accumulated to form the oceanic crust, which was known from earlier seismological studies to have an average thickness of about 5 km. Finally, because of the separating plates, the young crust of the ocean floor moves away from the ridges. We almost never see these eruptions directly, and hints about their occurrence were only uncovered in the late 1960s. The erupted material has come to be called MORB (mid-ocean ridge basalt). It is found beneath layers of various sediments that lie at the ocean bottom (average sediment thickness is roughly 2 km). To seismologists, the oceanic crust is defined to be part of the lithosphere even though it is a different material from the mantle. This crust has a small number of internal layers that form as the material solidifies at spreading centers. Those details will not concern us here. The bottom of the crust has been clearly known for a very long time because of a striking change in seismic velocity at its base. For our purposes, we are mostly interested in the flow of energy for crust formation, but let's start with an estimate of the volume flow rate of oceanic crust formation for Earth.

The total volume flow rate of the new oceanic crust created at spreading centers is rather easily estimated. We start with the area of the present ocean crust (this is easy to remember since the total area of all the ocean basins is 3.4×10^{14} m^2). To get a volume of crust at the present time, multiply the area of all the basins by the crustal thickness of (5 km) to get the ocean crust volume of 1.7×10^{18} m^3. The volume flow rate is equal to this volume divided by the duration of time since most basins have been created. This is roughly 200 million years (6.3×10^{15} s). This gives a volume flow rate of all oceanic crust accumulation at 269 m^3/s. This is the volume of the air within a rather large house that would be accumulated each second. It seems like a relatively modest number for our entire Earth, but it is a good example of how a strict adherence to the MKS system is not particularly useful for developing our visual picture. Let's "scale up" our imagination and calculate the volume flow rate based on the volume accumulated every million years, keeping in mind that the Earth is more than 3.5 billion years old. In 1 million years, the accumulated volume of oceanic crust is 8.5×10^{15} m^3. Therefore, we can consider that accumulation rate to be 8.5×10^{15} m^3/my (m^3 per million years). The units km^3/my are commonly used in Earth science so that rate is 8.5×10^6 km^3/my. Let's ask how much time it takes this volume flow rate to accumulate a volume equivalent to of all our continental material (called the "continental crust" by Earth scientists). The volume of all the continental crust is found using an area that is approximately half the area of the oceans (1/3 Earth's surface area) times a thickness of 36 km, so that the volume of continent material is 6.1×10^{18} m^3. The duration of time to accumulate a volume of ocean crust comparable to this is found by dividing 6.1×10^{18} m^3 by 8.5×10^{15} m^3/my = 717 million years. The oldest rocks of continental origin are approximately 3.6 billion years old, so 717 my is less than 1/5 the age of the Earth.

These simple calculations indicate that the magma accumulation rate in spreading centers is large enough so that if MORB were to be directly changed to continental crust, we would have much more continental material covering planet Earth than at present. Instead, much of the MORB must be recycled back into the mantle at subduction zones (There is more about the melting process that produces MORB in the mantle shortly).

The production of MORB is the largest volcanic process on Earth. Two other large sources deliver magma to the surface of Earth: island arc volcanos, and hotspots. The first of these, island arcs, are located at plate margins generally next to a subduction zone and a deep ocean trench. Many of them reside in the "ring of fire" around the Pacific basin (Fig. 4.1), but they also exist in other subduction zones that are near continents such as Greece, Italy, and the Caribbean. The magma feeding up into island arc volcanos is partly composed of materials associated with the subducted plate including old, weathered (chemically altered) MORB, continental crust, sediment, and water. In addition to having distinct differences in composition from MORB itself, the presence of water and carbon dioxide in the eruptions of arc volcanos tends to make them explosive.

To review, the three large magma sources are ridges, Island Arc volcanos, and hotspots (Fig. 4.2). The global total of the volume flow rates of all of them are not precisely determined because of many challenges to good data acquisition. First,

Fig. 4.1 Location of the ring of fire volcano belts around the Pacific Ocean. USGS public Domain

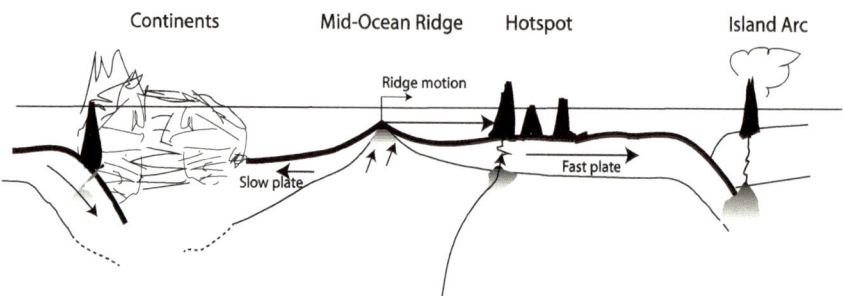

Fig. 4.2 Sketch of the three largest types of magma sources, ridges, arcs, and hotspots

some outpourings are explosive because of large amounts of water and carbon dioxide, so material can be cast in wide regions. Second, magma frozen around the flanks of these regions before arriving at the surface is hard to locate, so the volume of that frozen component is hard to estimate. Third, numerous smaller volcanos are not completely mapped. Fourth, to get a rate, the mass of each batch of material must be dated. However, crude estimates of the volume flow rates are still useful to consider.

Here, we attempt a crude estimate for the volume flow rate of island arc volcanism for the entire Earth. The detailed history of most volcanos is unknown, so this estimate is based on a detailed study along one string of volcanos. Detailed maps and age estimates are used for the 1100 km long Aleutian Island arc complex (Jicha et al. 2006). Their flow rates depend on ages based on radioisotope dating. The values of volume flow rate range from 89 to 182 km^3/km my, (the units are cubic km per million years per km of arc). To extend such an estimate to the entire world is of course quite reckless but never mind, let's proceed anyway. We use these values to adopt a typical value of 100 km^3/km my and extrapolate to the entire Earth by assuming the total length of arcs to be 10^4 km (but could perhaps be twice that). In total, the volume flow rate estimate for the volume flux rate of arc volcanos over the entire world is 31 m^3/s (10^{15} m^3/my or 10^6 km^3/my). This estimated rate is approximately 1/8 as large as the magma production flow rate of oceanic crust at spreading centers (8.5×10^6 km^3/my). The time for this flow rate to build continents is 6.1×10^{18} m^3 divided by 10^{15} m^3/my $= 6.1$ billion years.

Since MORB does not pile up and accumulate at ocean trenches, the subducted MORB must be recycled into the mantle, although it might contribute to a small percentage of island arc production. In addition, the volume flux estimate for island arc volcanos, if correct, indicates they are only marginally good candidates for adding significant material to the continents because it would take 6 billion years to gather the present volume of continental crust. This is longer than the age of the Earth of 4.6 billion years. More importantly, much of the chemical composition for island arc basalt requires compositional alterations to become continental material (granite). Therefore, both the rates and the chemistry imply that present island arcs are not candidates to build continents up to the present size.

Hotspots rank third in the size of the volume flow rate of magma that is delivered. What are hotspots? The story of how they came to be suggested to be a source of magma with deep mantle origins is a wonderful picture of deductive interpretation. Even before the great plates became known in early 1960, people had noticed the distinctive linear alignment of many island chains, principally in the Pacific Ocean but to a lesser extent in other locations worldwide, too. The Hawaiian Island chain is the clearest example. In 1963, the Canadian Geologist Tuzo Wilson suggested that such volcanic island chains may have formed due to the movement of a plate over a stationary "hotspot" in the mantle. His idea was considered so radical that his "hotspot" manuscript was rejected by many major international scientific journals and was ultimately published in the Canadian Journal of Physics. He pointed out that Hawaii has active volcanism in the largest, youngest island, and the older islands extend along a linear trend (on the spherical Earth, of course) and they become progressively older, even to the extent of the older ones being eroded completely from the ocean surface and classified as seamounts. Wilson mentioned other linear trending island chains, too.

There were numerous competing ideas about the mechanisms generating the linear chains. Their locations are not correlated with any locations on the plates, and their

Fig. 4.3 A collection of
hotspot products

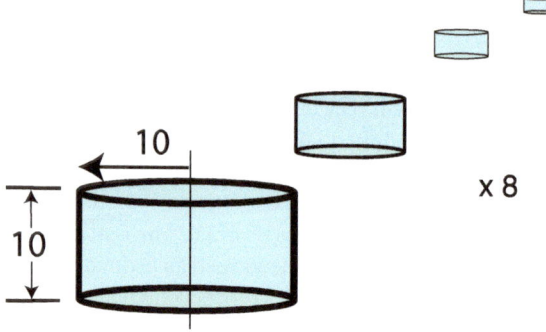

composition has significant differences from both MORB and island arc basalts. Many people suggested a giant crack, compositional weaknesses, or simply material that melts easily to make the outbreak of the volcanos in the island chains. These slowly gave way to the hotspot hypothesis as data revealed several features below hotspots consistent with both a fixed source of heat (a source of heat flow) and even a hot column of mantle rising beneath each hotspot. The roots of both Iceland and Hawaii are detected seismically to many hundreds of kilometers below the surface. Trace elements in the lavas have distinctly different patterns from MORB, so the possibility that hotspots come from very deep in the mantle is actively supported. Here, we estimate the total volume flow rate for hotspot magma.

To estimate the rate of delivery of magma to the surface for hotspot volcanos, we take a typical volcano on a hotspot island to have a volume equal to that of a cylinder 10 km in radius and 10 km high and take an eruption rate of every million years. This gives a rate of 0.1 m³/s. Then, let's take as a model a total of 8 chains (Fig. 4.3) to get a volume flow rate for island chain volcanos. This comes out to a delivery rate of magma to the surface of 0.8 m³/s.

4.2   Flow Localization in the Laboratory

I was fortunate to be able to conduct some laboratory experiments and theory in the period 1969–1973 that showed the structure of rising viscous liquid when it flows up from a deep source. My motivation was purely from a fluid mechanics point of view. I wanted to find out what happens when viscosity contrasts are very large. However, the original suggestion was geological and concerned salt-domes, which are large uplifting layers of salt in sedimentary basins (The site of many oil wells, by the way). After a lecture at UCLA about the salt-domes by visitor Bill Chappel from Brown University, David Griggs asked me whether we had two fluids of high viscosity in the lab that don't dissolve together. We could put them in a cylindrical container with a thin top layer of the lighter fluid and the other fluid in the remainder of the tank.

We slid a plexiglass plate on top to exclude air, carefully leveled the plate, left it overnight, and inverted it all the next morning.

If we take two fluids with different densities and viscosities in a field of gravity and ask about the form of motion as the lighter fluid rises and the denser fluid sinks, we expect that the answer will depend on which fluid is lowest in viscosity and which layer is deepest. Although it is reasonable to expect that the low-viscosity fluid will be more mobile and therefore adopt a finger-like shape compared to the higher viscosity fluid, the flows in my first experiment did not confirm this expectation. Therefore, two sets of experiments were systematically conducted to show what occurs. They were combined with a linear instability analysis like the analysis by Rayleigh, but in this case, there were two layers of motionless fluids having a tiny perturbation. The theory predicted the wavelength and rate of growth of instabilities (Whitehead and Luther 1975).

The first set of systematic experiments used 2 transparent boxes containing a very viscous silicon oil and glycerin with differing thicknesses. The oil's viscosity was 44 times greater than glycerin viscosity. One box had a thin layer of the denser glycerin lying at the bottom of a deeper layer of more viscous silicon oil. The second box had the thicknesses reversed so that the thin layer of silicon oil was above the deep layer of glycerin. (The Fig. 4.6 caption clarifies the actual experiment). The boxes had rigid, sealed lids with no air, and they were very carefully leveled and left overnight or longer. To observe the evolution, because the two fluids were viscous, we could quickly invert the boxes and photograph the evolution of the interface as the light fluid rose and the dense fluid sank over the next few minutes. In both boxes, soon after inversion, the fluid pushed out of the thin layer as circular protrusions. The centers of Benard's hexagonal flows come to mind as in Fig. 4.4.

Here are the results. With the thin layer more viscous as in Fig. 4.5, then after inversion, the thin layer of more viscous but lighter silicon oil was below the glycerin. Spacing between the protrusions was about the same distance as the original layer thickness. The rising columns of higher viscosity silicon oil became vertically elongated, and their internal forces were clearly dominated by a balance between buoyancy and internal stress within the column. In contrast, with a thin layer that

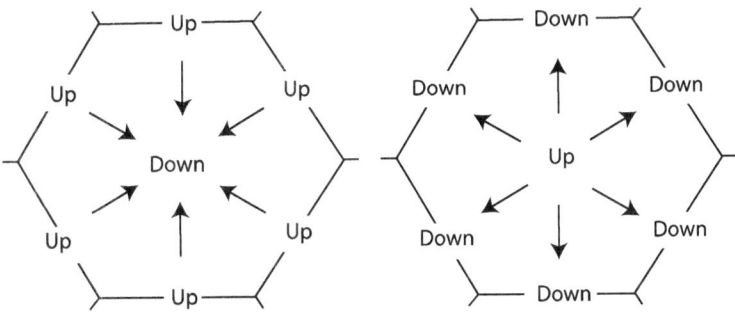

Fig. 4.4 The region and ascending and descending flow for the hexagon planform

Fig. 4.5 A thin layer of relatively high viscosity silicon oil rising through transparent glycerin (Whitehead and Luther 1975)

was less viscous (thin layer of glycerin starting above silicon oil and sinking down), the spacing was much greater. Figure 4.6 shows what happens with a low viscosity thin layer. Note that the photo is inverted to help visualize a layer of salt rising. The internal forces for protrusions of lower viscosity fluid were clearly dominated by a balance between buoyancy and stress of the viscous host fluid. The centers of fluid pushing out from the thin layer of lower viscosity fluid became gathered in large spherical-like globs that were aided by lateral flow in the thin layer to gather the fluid together. The spherical-like globs slowly sank down (they appear to rise in Fig. 4.6, which is inverted). Later, a very thin vertical conduit with internal pipe flow trailed each spherical-like glob. It contained small amounts of remnant fluid that had not been drained by the great gathering event.

The second set of experiments showed how the flow evolved if a fluid source continues at a steady rate. To see this, we used two oils of similar composition but different viscosity (More recent experiments use simple corn syrup with the second liquid or the host fluid having just a small amount of water added for viscosity and density contrast or only a temperature difference). With the syrup, if a higher viscosity fluid at the top of lower density fluid is steadily pumped in, we saw vertical columns of viscous fluid descending through the less viscous fluid. The columns were very similar to the ones in the first set (Fig. 4.5). If a lower viscosity fluid was steadily pumped into the bottom of higher density fluid, the fluid first formed a roughly spherical-like glob that slowly moved up as in the first set (Fig. 4.7). However, since we continued to inject more fluid at a steady rate, the spherical-like cavity was followed by a small column of flowing fluid that allows a relatively large flow rate of material to continue flowing up through a small circular cross section (like a conduit, or pipe) for as long as pumping continued. This indicates that the most rapid vertical motion of all occurs within the lower viscosity fluid in a vertical tubular pipe.

Pumping the fluid in different ways reinforced the fact that the lower viscosity fluid would start up as a relatively large spherical-like glob. Figure 4.8 shows a linear

Fig. 4.6 A thin layer of relatively low-viscosity glycerin (dyed) moving vertically through transparent silicon oil. In practice, since glycerin is denser than the oil, the dyed layer was along the top and we photographed up into the layer. The photo is inverted to make it more familiar to Earth scientists (Whitehead and Luther 1975)

row of upwellings in a lower viscosity material pumped in along a horizontal line. Figure 4.9 shows what happens when the line is tilted at an angle to the horizontal. In both cases, the upward projections start with spherical-like globs.

If the source of low-viscosity fluid into a vertical pipe is unsteady in time, the pipe flow responds with waves flowing upward. This new feature is shown in Figs. 4.10 and 4.11. The solitary waves pass through each other during a collision (Scott et al. 1986). Similar results were discovered independently by Peter Olson at about the same time. Figure 4.11 shows how the waves convey trapped material upward with interior recirculation and thus can convey material over long distances (Whitehead and Helfrich 1988).

The possibility was immediately obvious that any flow within the mantle of the Earth like this conduit might support the flow of warmer, lower viscosity deep mantle material upward to feed hotspots. The experiments suggested that hotspot material might rise from some layer in the deep mantle where the hot material accumulates so that circular upwelling forms.

These flows within conduits proved to be also good models of the flow of liquid in a porous matrix next to crystals in the mantle. Even the waves are similar. Therefore, the results and associated theory help to build a fundament understanding of differential flows between two fluids inside the mantle. Another fundamental result

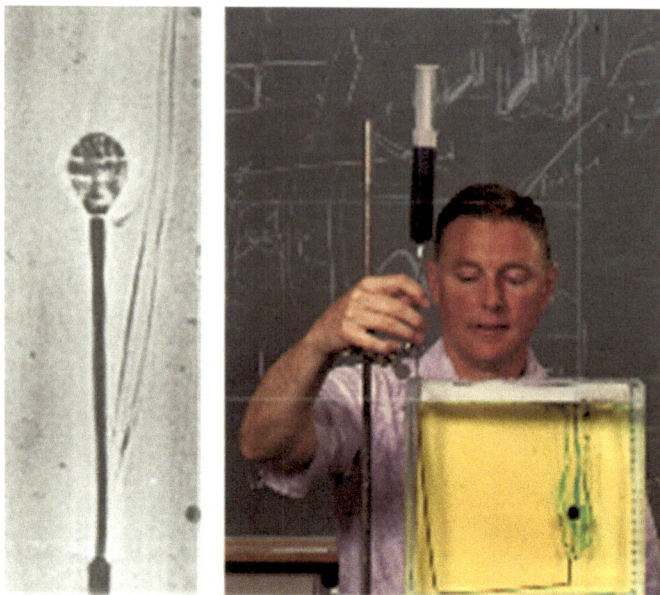

Fig. 4.7 (Left) Image of rising flow of fluid of lower viscosity and density in a laboratory experiment in 1973 (Whitehead and Luther 1975). The dark fluid rises through a denser fluid with 6000 times higher viscosity. The initial body is close to a sphere and the trailing body is laminar pipe flow. (Right) A simple classroom demonstration using corn syrup. The pumped corn syrup has a small amount of water and dye mixed in

Fig. 4.8 A line of lower density and lower viscosity fluid (corn syrup with a bit of water added) injected at uniform volume flow rate by a laterally moving syringe into corn syrup. It develops instabilities with miniature plume heads and tails (Whitehead et al. 1984)

of the experimental flows is that they never demonstrated rising sheet-like structures. In fact, it is interesting to compare these flows with the spoke convection cells described in the previous chapter. Spokes are also circular concentrated regions of vertical flow in the interior. The actual spokes are instabilities in the boundary layers in localized flow that radially converges into the vertical flow regions. Both

Fig. 4.9 Injecting a vertical conduit and then tilting it produces instability that creates new plume heads Skilbeck and Whitehead (1978)

Fig. 4.10 A large solitary wave overtaking a smaller one, passing through it, and leaving it behind. This figure shows that there is some mixing of material during the interaction

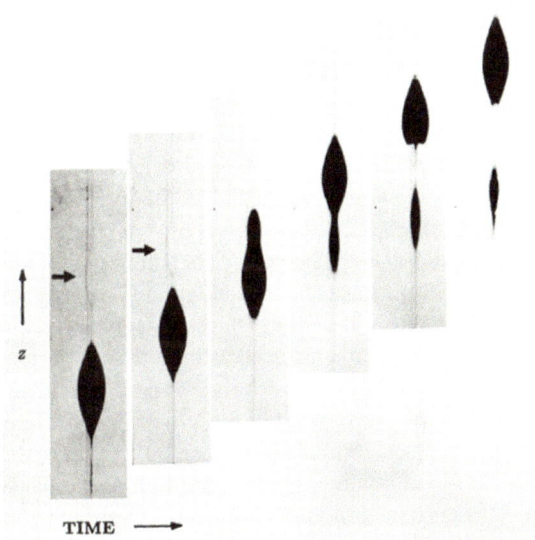

TIME ⟶

Figs. 4.5, 4.6 and 4.7 and the spokes seem to eliminate sheet-like flows as possible flows in large Ra convection with uniform viscosity. In summary, from the perspective of fluid mechanics, these experiments and theory give good reasons to expect rising or sinking circular columns within the mantle.

The flows in the laboratory experiments and theory sharply break away from the structure and behavior of the roll cells described in Chap. 2. In fact, at first, the referees of an early version of my submitted manuscript could not see my point. I had to revise the paper extensively with Doug Luther to make the results sensible in a geological or geophysical context. Fortunately, there were so few hints in the

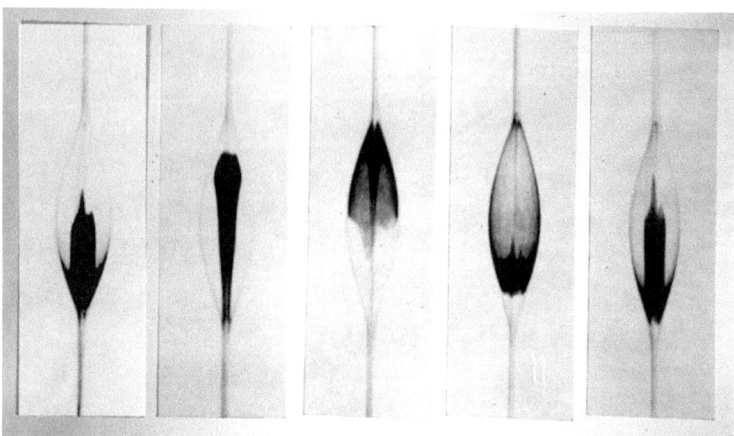

Fig. 4.11 Closed circulation within a moving wave (Whitehead and Helfrich 1988)

literature about what sort of motion might be expected both in salt-domes and under hotspots that the laboratory results and supporting theoretical understanding of the mechanics have proved to be useful.

The experiments in Figs. 4.5, 4.6, 4.7, 4.8, 4.9, 4.10 and 4.11 concern the rise of material of different densities and viscosities in an ambient fluid. No melting or magma production is included, but the experiments led the way to understanding liquid migration. How does the arrival of material from greater depth in the Earth produce volcanism? The answer is simple. First, when mantle material rises, it is below the Earth surface and each grain is moving up to lower pressure. Second, in our present Earth, the upper mantle is evidently composed so that there is a slight melting of the mantle material as the pressure decreases and moves below a certain value. The first liquid to form is composed of those elements that have the lowest melting temperature, notably those with very different molecular weights than the bulk mantle elements. These include both water with the very low molecular weight of hydrogen and uranium with very high molecular weight. We explained earlier in this chapter that if the liquid spreads out around some of the solid mantle crystals and conditions allow the liquid around the crystals to connect (a function of both the surface tension of the melt-crystal interfaces and of the relative volume of the liquid compared to the volume of the solid). The liquid will flow up if it has a lower density than the crystals or down if it has a higher density. I like to visualize that the solid crystals slowly deform (by the movement of dislocations in the crystal structures) and "squeeze" the liquid up or down, just like squeezing water out of a sponge. In the Earth, the liquid that is lower in density than the mantle material flows upward and serves as the primitive source of magma under the mid-ocean ridges and volcanos. Chemically, the rising liquid has a different composition from the mantle. Further, it alters its composition by reacting with host materials as it flows up to lower pressure. Finally, the rising fluid arrives near the top and is cooled. The liquid composition changes here as well by alteration of the chemical composition of the magma by crystal formation before solidification is complete (Jackson et al. 2018). Inferring the very complicated chemical history of solidified lava in surface rocks that were altered by contact with magma is the realm of geochemists and their fascinating and challenging science. I cannot explain the complexities of their stories here (in fact, some still confuse me). For purposes of this book, the energetics and fluid flow dynamics can be simplified to a few calculus equations, although taking these equations and finding specific mathematical solutions that express the flows is not always easy.

Fortunately, there are some analogs available in laboratories to help us. For example, a simple model of the porous flow in a matrix turned out to be modeled by the conduit flow in Fig. 4.7. This thought came to me as a "eureka" moment as I was listening to a lecture by Frank Richter on porous flow as he expressed his desire to find a simple fluid model. I pictured the liquid in the mantle being like the fluid within the circular conduit. The crystals in the mantle and their creep act like the surrounding more viscous fluid. It was obvious that the mathematics of the two problems were very similar. I ran to the lab to clarify my thinking. This led to experiments and theoretical work with a variety of people since then. Overall, we have found that porous flow and the circular conduit have a variety of similar features including some distinctive waves that are like those in Fig. 4.10 that might rise within the mantle.

I worked over a span of 30 years on some interesting flows with geological applications that occur when the viscosity contrast is large. In terms of the flow of energy, rising mantle crystals in the Earth not only bring hot material upward but also begin to melt when the pressure gets below a certain value that depends on their composition. The melting uses some of the energy stored as heat within the material to convert the solid to a liquid. As mentioned already, the liquid collects either within an interwoven network around the crystals or in drops, depending on the surface energy between the liquid and crystal. The networked liquid rises, releasing potential energy. This energy release is balanced by the energy consumed by frictional drag. The result is that hot magma pours out at the surface and releases great heat in very localized eruptive regions. The eruptions are universally unsteady.

This vertical movement of magma has produced the largest separation of materials within the Earth by leading to the formation of the Earth's crust. Generally, the relatively uniform upward porous flow of liquid is focused into concentrated regions, either because of instabilities shown in Fig. 4.8, or because of fingering of reactive liquid flowing through pores. This is demonstrated in the simple experiment shown in Fig. 4.12 showing a side view of a mixture of glass balls and salt in a narrow slot. It is an inverted model of the upward flow of liquid through soluble crystals. The bottom of the slot allows water to drain uniformly through a sponge. Before starting, water is filled into the tank with the glass balls and salt present. After a short time, the water in the mixture dissolves enough salt to be saturated. There is still enough salt, so some is left in the glass matrix. The saturated water lying in the salt-glass ball matrix is much denser than the fresh water above the mixture with no circulation between the mixture and above. The experiment starts when water is slowly fed from above, and the same value of volume flow leaves through the porous bottom. Gradually, channels form where salt is dissolved away and only the balls remain. In

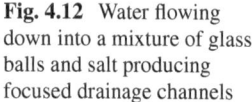

Fig. 4.12 Water flowing down into a mixture of glass balls and salt producing focused drainage channels

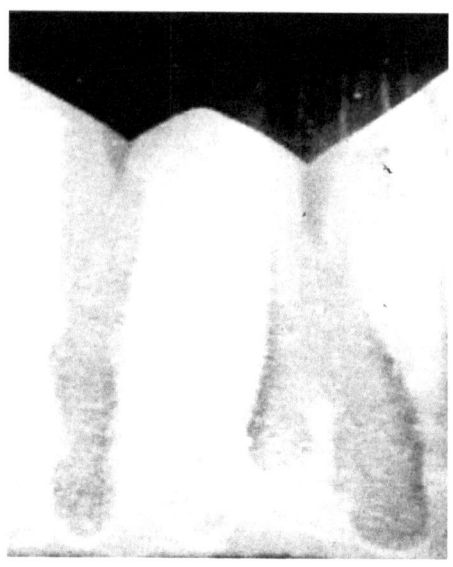

addition, the top of the mixture sinks over the channels. Both the sinking of the top and the dissolved channels produce paths of low flow resistance that focuses almost all the draining water into the channels. This focusing exists even though buoyancy of the saturated water would produce a uniform flow down. In addition, the bottom sponge would produce uniform flow down, so both stabilize the focusing with the presence of dissolution. Theory and numerical models of flow in compaction driven flow predict similar channels (Aharonov et al. 1995; Kelemen et al. 1995). The theory in these models, and previous work by others, found waves like those in Fig. 4.10.

The localization of flow of liquid can also be caused by cooling and solidification. Paraffin spreading out radially in a slot from a point source over a cold metal block forms localized channels of flow (Fig. 4.13).

4.3   Arrivals and Eruptions, Flood Basalts, and Extinction Events

The experimental photos of vertical motion that are shown in Figs.4.5, 4.6, 4.7, 4.8, 4.9, 4.10, 4.11 and 4.12 were made decades ago. Since then, numerous other experiments and the development of numerical models have led to many additional aspects of the flow dynamics, but the basic forms of flow are like the ones reviewed here. First, upwelling regions tend to be localized and circular rather than sheet-like. Second, a large initial 'head' is often present when upward flows of lower or equal viscosity are suddenly started compared to the ambient fluid. Third, there is only a small head for upwellings of higher viscosity compared to the surrounding fluid.

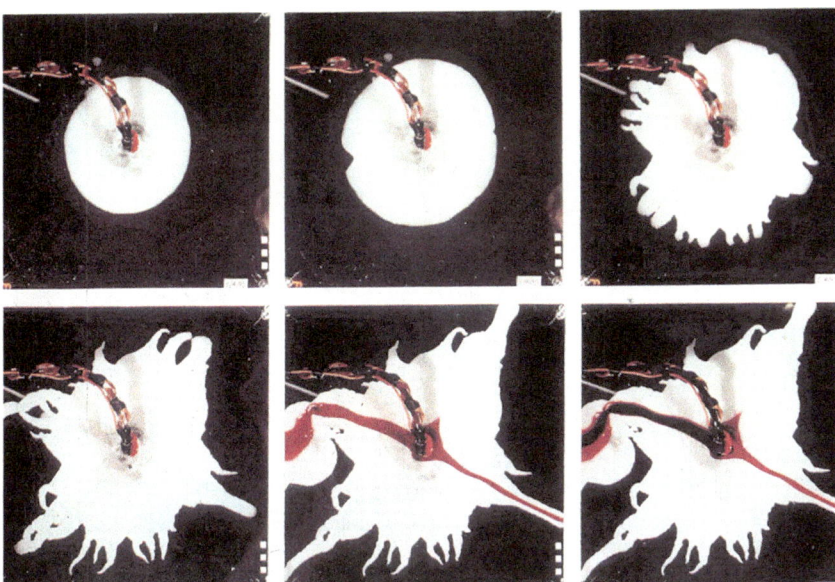

Fig. 4.13 Paraffin spreading in a slot with a transparent upper surface over a plate that is kept at below freezing temperature. It spreads evenly at first in a radial direction, and then focuses into many drainage channels (top, right and bottom, left), then 2 channels indicated by red dye (middle bottom), and then 1 channel indicated with black dye (right bottom). All the old channels are frozen shut. Plate 1 in Whitehead and Helfrich (1991)

There is seismic evidence for concentrated hot regions under Hawaii (Li et al. 2000; Huckfeldt et al. 2013) and Iceland (Wolfe et al. 1997) and one might presume these are like the low-viscosity conduits in the laboratory. Another exciting application concerns flood basalt events. These are massive eruption events that occur at the beginning of a linear hotspot chain. They are well-known to geologists, because some of the massive eruptions have occurred at the same time as large extinction events. Recent evidence indicates that they were large enough to alter climate and thereby provide a mechanism leading to the extinctions. The arrival of the large spherical-like cavities at the start of our experiment can be used as a model of the arrival of such large cavities of hot mantle material from deep in the mantle to Earth's surface leading to massive eruptions of magma to produce flood basalts. The trailing conduit, possibly tilted as in Fig. 4.9 by mantle shear, has laid down the rest of the chain. Figure 4.14 shows some of the well-known larger flood basalt locations, and some of the trails of smaller volcanos aligned as the surface plate moves.

The biggest contribution of magma flows to the surface of the Earth is to form crust, and there are two different types. One is the oceanic crust, which is typically 5 km thick. Almost all of it has been laid down by cooling magma at spreading centers and has accumulated during the age of present plates, which is approximately 200 million years. A century ago, the magma flows that supply this crust were unknown but were

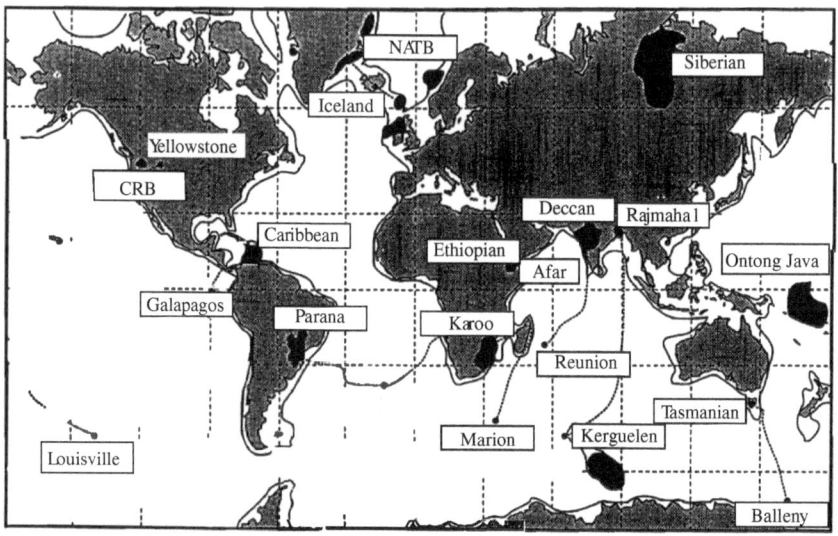

Fig. 4.14 Flood basalt locations erupted over the last 250 million years that might be produced by the arrival of a plume head (Richards et al. 1989)

suspected to be volcanic because of trawl samples of weathered lava material. The spreading center origin was discovered in the late 1960s when concentrated lava formations were discovered at the central valleys of mid-ocean spreading centers using camera pictures and visual reports in the submarine Alvin. The evidence for volcanism at mid-ocean ridges was circumstantial at first. People identified forms of the lava flow, but soon geochemistry supplied compelling evidence that the ocean crust recovered by dredges was like weathered lava. The formation of ocean crust with all its complexity is a large field of study for marine geologists. Although cooled magma is common in ridge areas, only recently has an actual magma outflow at a ridge been documented on the southern Juan de Fuca Ridge between 1981 and 1987 (Chadwick et al. 1991).

The continental crust, in distinct contrast, is much older and of a much different composition with a lower density. Its average thickness is estimated to be 36 km. The history of the material is very much older, complex, and different at each location. It is formed by modifying ocean crust that has been subducted and mixed with sediment and water. Other examples have modifications in the presence of older continental crust, as well. The history of each sample differs, with some crust being solidified few 100 million years ago and other crust solidified over 4 billion years ago. There is evidence that some continental crust is forming even now.

Other features on the surface of Earth might be connected to convection cells within the mantle as well. Figure 4.14 shows the results of a recent numerical study of convection with uniform viscosity heated from below by Lees et al. (2020) that takes the form of spoke convection. It suggests that spoke flows might occur in the mantle under Africa. Busse and I (1974) reported that we found the spoke pattern in laboratory experiments, and I am both gratified and excited to see this recent work as I write this 50 years later. These numerical calculations were not possible then. Their analysis is useful in ways not suggested by our laboratory observations because they can compare flows with different types of top and bottom boundaries. For example, the numerical code can readily have free-slip conditions at the top and bottom, so fluid moves along the boundary. Most laboratory studies have rigid boundaries so the movement at the boundary is zero. The numerical study that has produced the spoke pattern in Fig. 4.15 shows that many different possible mechanical boundary conditions along the top and bottom surfaces of a layer of convection still make the spoke pattern for Ra greater than 10^5. This adds to the laboratory evidence that the spoke pattern is quite robust and likely to be expected. In addition, Lees et al. (2020) produce a variety of geophysical measurements and compared them with actual measurements. The numerical results produce an estimate of magma production from small volcanos in Africa (Fig. 4.16). Their comparisons between computation and field data (Fig. 4.17) give evidence that ordinary convection exists below the African Continent.

4.4   Energy-Flow Rates of Mantle Convection

Returning to the topic of mantle convection, numerical models have been developed by several groups around the world. It is difficult to say which are best, since many of them are created for specific objectives. All of them are hampered because they need to include several physical factors that are poorly known. The distribution of both radioactive heat production and viscosity within plates and under volcanos is still uncertain in detail. Even the concept of a viscosity might become invalid at the edge of the plates and within the continents. Also, a lack of knowledge of the details of plate and continent evolution back in time greater than a few 100 million years make it hard to know which model works best. Despite these challenges, the numerical models of mantle convection impressively have overcome some challenges such as taking account of internal phase changes within the mantle as they march forward in computer time to show some aspects of the evolution and the structure of the Earth.

Estimates of flow rates help us to understand the relative size of the motions driven by energy flow within the mantle. We begin with an estimate of the volume flow rate to supply all the plates so it can be compared to the rate of magma supply. The overall dimensions of the collected plates are represented as the one giant long plate sketched in Fig. 4.18. We use the average plate motion speed: $u = 10^{-9}$ m/s, the total length of spreading centers $l = 5.5 \times 10^7$ m, and an average plate thickness of 100 km or $h = 10^5$ m to give a volume flow rate of $Q = l \times u \times h = 5.5 \times 10^3$ m^3/s.

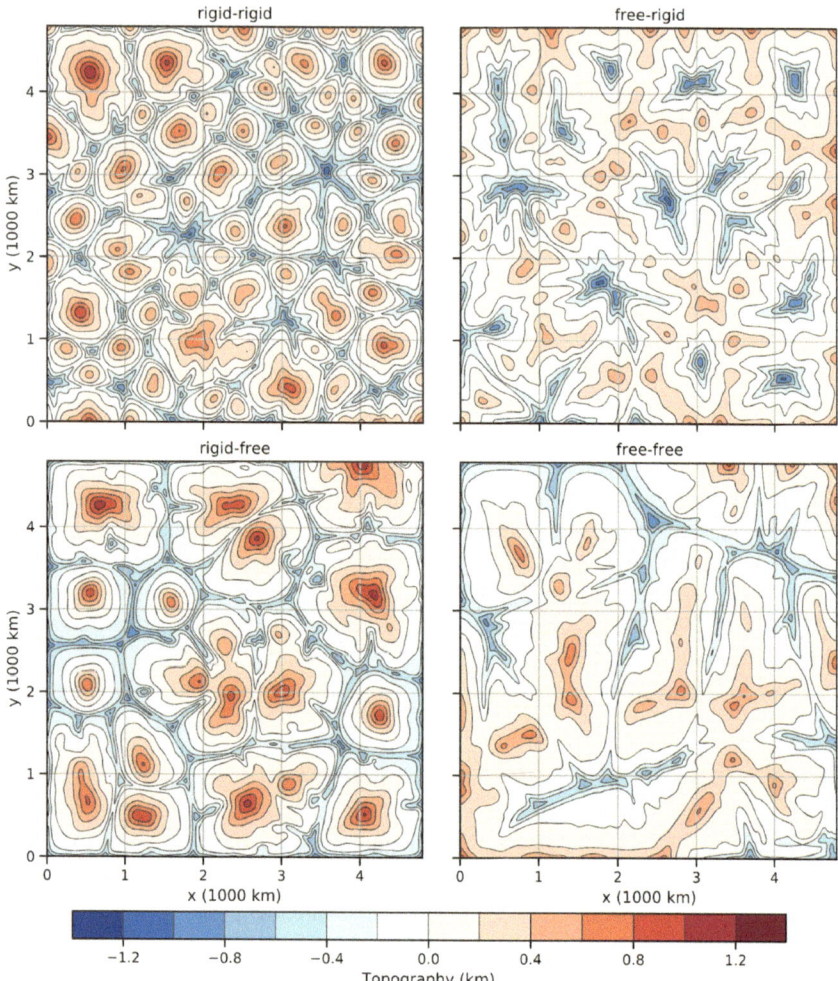

Fig. 4.15 Numerical calculations for surface elevation at the top of a layer of mantle under Africa. Ra = 10^6. Figure 4 in Lees et al. (2020) used under Creative Commons license

It is useful at this stage to make a table showing simple estimates of volume flow, buoyancy, and heat flows for five surface features of Earth. The table has three rows that show: 1. The volume flow rate. 2. The buoyancy flow rate, (the buoyancy force times the volume flow rate). 3. The heat flow rate delivered from the interior of the Earth to the surface. The columns are the plates, the ocean crust, hotspot upwelling, island arc volcanos, and hotspot volcanos. Each estimate in Table 4.1 is described in Sect. 4.5.

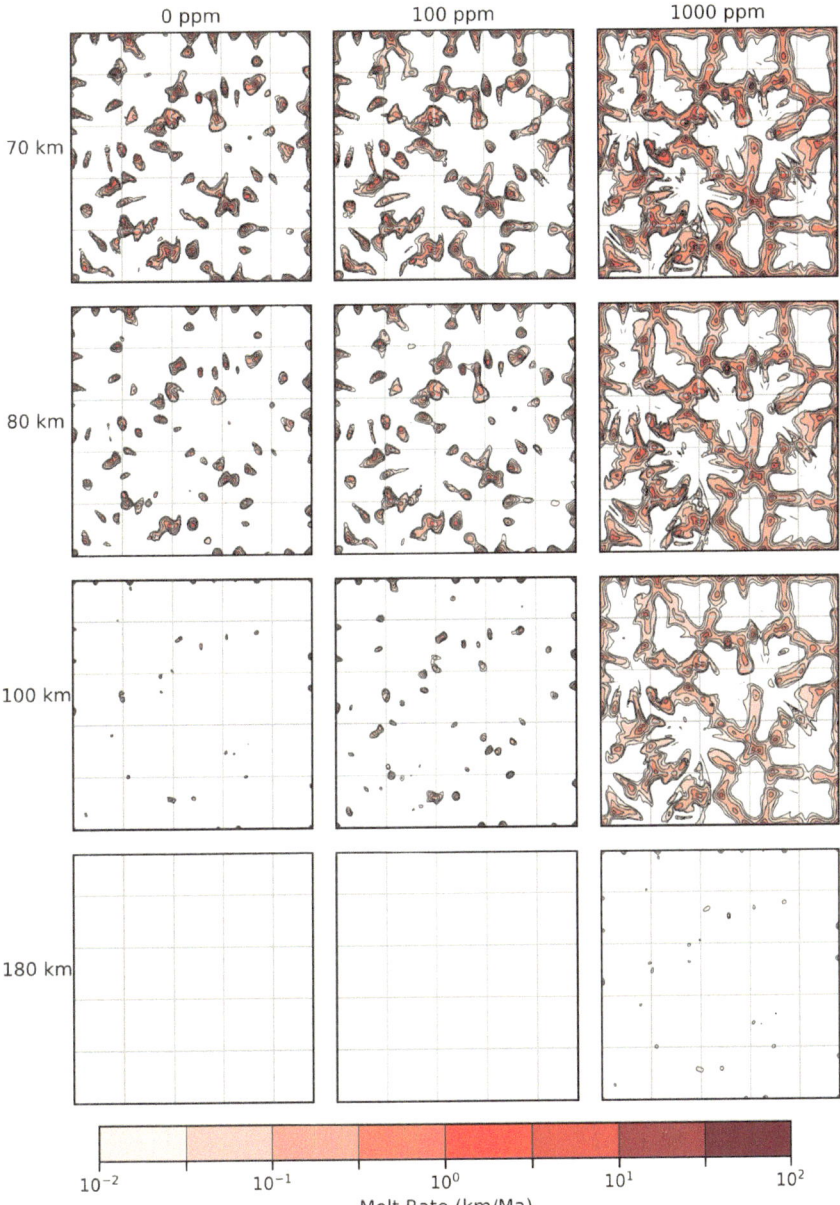

Fig. 4.16 An estimate of the amount of magma produced at the top of the convection cells in Fig. 4.15. Figure 10 in Lees et al. (2020) used under Creative Commons license

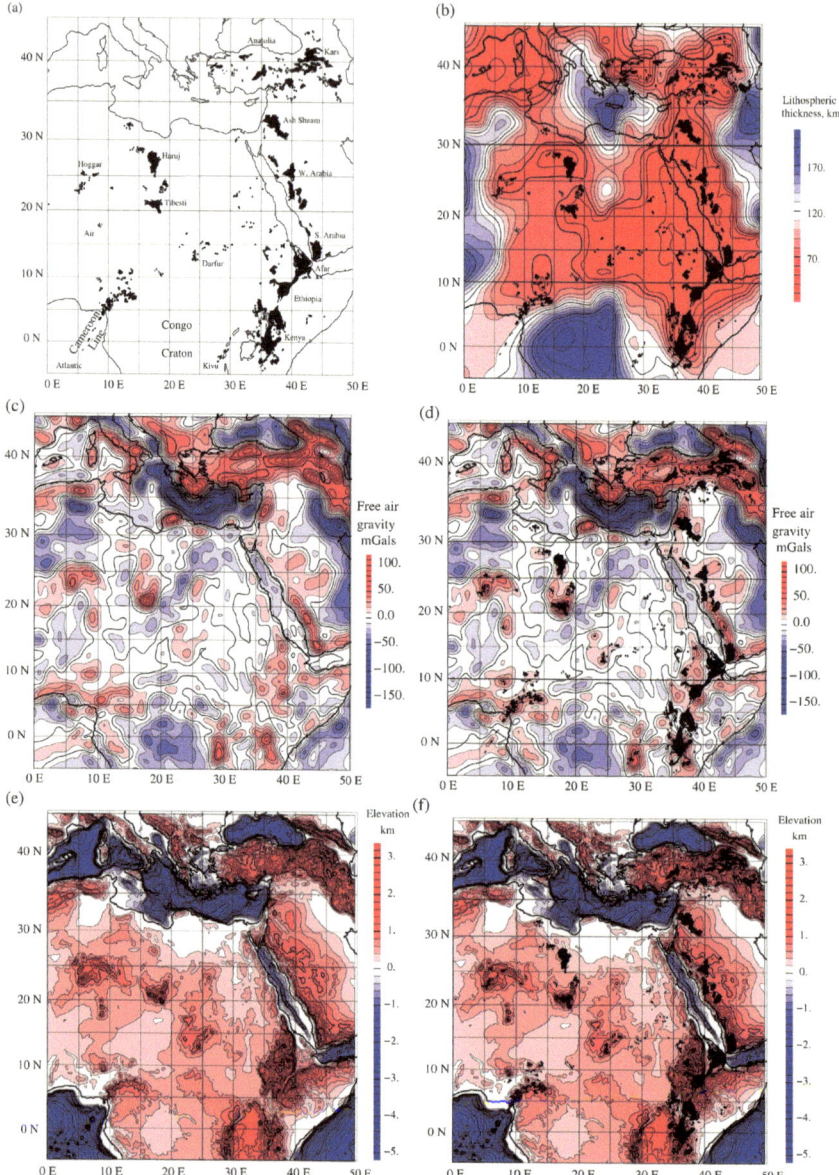

Fig. 4.17 Lithosphere thickness, gravity data, and surface elevation for the portion of Africa in Fig. 13 in Lees et al. (2020) used under Creative Commons license

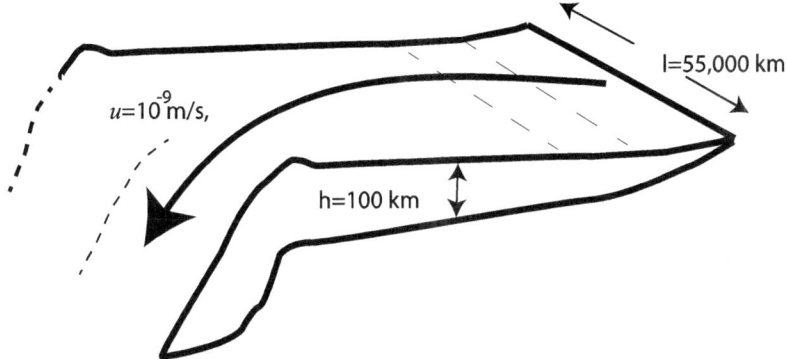

Fig. 4.18 Sketch of the sizes used to estimate the rate of volume flow for plate formation

Table 4.1 Some components contributing to the energy flow of the Earth. As a comparison, the present worldwide human energy flow based on consumed fuels is approximately 2×10^{13} W and the total sunlight intercepted by Earth (total radiation, no reflection) is 6×10^{17} W

	1. Plates	2. Ocean crust	3. Hotspot upwelling	4. Island arc volcanos	5. Hotspot volcanos
Volume flow rate Q (m³/s)	5500	269	30	31	3
Buoyancy flow rate b (n/s)	7.76×10^6	7.5×10^6	4230	8.7×10^5	8.4×10^4
Heat flow rate (W)	2.6×10^{13}	5.1×10^{11}	1.4×10^{10}	5.8×10^{10}	5.6×10^9

4.5 The Calculations

Row 1 has estimates of the volume flow rates (given by the letter Q) associated with 5 aspects of mantle convection and magma production. The first column is sketched in the production-subduction cycle of a plate in Fig. 4.18 and is for all the lithospheric plates. Column 3 in row 1 is a volume flow rate for the rising within the mantle that feeds hotspots. For this, I decided to rely upon the estimate by Davies (2001, p. 298) (That chapter covers many of the topics in this chapter, too). He uses the width of the Hawaiian swell that extends around the Hawaiian hotspot chain on the Pacific plate and gives a value of volume flow rate of 3 m³/s for the upwelling of mantle material under Hawaii. Since we want a value for the entire Earth, and since the Hawaiian hotspot is clearly much larger than any of the other 40 hotspots, we multiply it by 10, so the flow is 30 m³/s. Columns 2, 4, and 5 are values of the volume flow rate of melted liquid accumulation calculated earlier in this chapter.

Row 2 has estimates of the rate of production of buoyant forces associated with each item. It expresses the aggregate buoyancy within Earth and is given by the formula $b = g * \rho_0 * \alpha * \Delta T * Q$ with the units of Force per second (n/s). Column 1 is for subducting plates. The value used for gravity acceleration is rounded off to $g = 10$ m/s^2, while density is $\rho_0 = 4700$ kg/m^3, the coefficient of thermal expansion is $\alpha = 3 \times 10^{-5}$, and the temperature difference between the upper mantle and cold slabs is $\Delta T = 1000$ °C. For column 3, we use $g = 10$ m/s^2, $\rho_0 = 4700$ kg/m^3, $\alpha = 3 \times 10^{-5}$, and $\Delta T = 100$ °C for supplying the Hawaiian swell. Columns 2, 4, and 5, for the magma buoyancy forces, flow rates for ocean crust formation, island arcs, and hotspot volcanos are given by the simple formula $b = g \times \Delta\rho \times Q$ where the density difference between liquid and solid is approximately 2800 kg/m^3.

Row 3 is a heat flow rate. The formula for columns 1 and 3 (plates and hotspot upwelling) is $H = \rho_0 \times C_p \times \Delta T \times Q$. Column 1 uses density $\rho_0 = 4700$ kg/m^3, $C_p = 1000$ j/kg °C, a temperature difference of $\Delta T = 1000$ °C and $Q = 5500$ m^3/s. Column 3 uses density $\rho_0 = 4700$ kg/m^3, $C_p = 1000$ j/kg °C, and $\Delta T = 100$ °C. Columns 2, 4, and 5, use the formula $H = \rho_0 \times L_a \times Q$ with a value of latent heat $L_a = 4 \times 10^5$ j/kg instead of $C_p \Delta T$ (equivalent to the heat needed for a temperature change of 400 °C).

In conclusion for this chapter, we show that there are several flows of material and heat within the mantle that are believed to lead to magma production in the Earth. The magma centers generally tend to cluster into concentrated upwelling regions that in many cases are cellular in form and are accompanied by a significant flow of energy up to the surface. They are always unsteady. Heat flow is concentrated near the upwelling locations, and the production of melted magma is the consequence of this concentrated upward flow. The energy-flow rate for magma production is generally lower than for the major plates and mantle convection, but the chemical differentiation from the associated magma formation is large and leads to the crustal material that supports life on Earth.

References

Aharonov E, Whitehead JA, Kelemen PB, Spiegelman M (1995) Channeling instability of upwelling melt in the mantle. J Geophys Res: Solid Earth 100(B10):20433–20450

Busse FH, Whitehead JA (1974) Oscillatory and collective instabilities in large Prandtl number convection. J Fluid Mech 66(1):67–79

Chadwick WW Jr, Embley RW, Fox CG (1991) Evidence for volcanic eruption on the southern Juan de Fuca Ridge between 1981 and 1987. Nature 350(6317):416–418

Davies GF (2001) Dynamic Earth: plates, plumes and mantle convection. Cambridge University Press

Huckfeldt M, Courtier AM, Leahy GM (2013) Implications for the origin of Hawaiian volcanism from a converted wave analysis of the mantle transition zone. Earth Planet Sci Lett 373:194–204

Jackson MD, Blundy J, Sparks RSJ (2018) Chemical differentiation, cold storage and remobilization of magma in the Earth's crust. Nature 564(7736):405–409

Jicha BR, Scholl DW, Singer BS, Yogodzinski GM, Kay SM (2006) Revised age of Aleutian Island Arc formation implies high rate of magma production. Geology 34(8):661–664

Kelemen PB, Whitehead JA, Aharonov E, Jordahl KA (1995) Experiments on flow focusing in soluble porous media, with applications to melt extraction from the mantle. J Geophys Res: Solid Earth 100(B1):475–496

Lees ME, Rudge JF, McKenzie D (2020) Gravity, topography, and melt generation rates from simple 3-D models of mantle convection. Geochem Geophys Geosyst 21(4):e2019GC008809

Li X, Kind R, Priestley K, Sobolev SV, Tilmann F, Yuan X, Weber M (2000) Mapping the Hawaiian plume conduit with converted seismic waves. Nature 405(6789):938–941

Richards MA, Duncan RA, Courtillot VE (1989) Flood basalts and hot-spot tracks: plume heads and tails. Science 246(4926):103–107

Scott DR, Stevenson DJ, Whitehead JA Jr (1986) Observations of solitary waves in a viscously deformable pipe. Nature 319(6056):759–761

Skilbeck JN, Whitehead JA Jr (1978) Formation of discrete islands in linear island chains. Nature 272(5653):499–501

Toomey DR, Purdy GM, Solomon SC, Wilcock WS (1990) The three-dimensional seismic velocity structure of the East Pacific Rise near latitude 9 30′ N. Nature 347(6294):639–645. https://doi.org/10.1038/347639a0

Whitehead JA Jr, Luther DS (1975) Dynamics of laboratory diapir and plume models. J Geophys Res 80(5):705–717

Whitehead JA, Helfrich KR (1988) Wave transport of deep mantle material. Nature 336(6194):59–61

Whitehead JA, Helfrich KR (1991) Instability of flow with temperature-dependent viscosity: a model of magma dynamics. J Geophys Res 96(B3):4145–4155

Whitehead JA Jr, Dick HJ, Schouten H (1984) A mechanism for magmatic accretion under spreading centres. Nature 312(5990):146–148

Wolfe CJ, Bjarnason I, VanDecar JC, Solomon SC (1997) Seismic structure of the Iceland mantle plume. Nature 385(6613):245–247

Chapter 5
Power and the Building of Continents, Mountains, and the Ocean Basins

Abstract The distribution of continental crust has evolved so that the continents and ocean basins form their present structure. Mantle convection produces plates that form at ridges and vanish at subduction zones, so continents are steadily conveyed to subduction zones. There, the continental crust is thickened by lateral compression and mountain building. This is balanced by erosion of the continent surface. The result is that continent elevation has a small value above the ocean surface compared to its thickness, and this makes the ocean basins form a cistern with exactly the correct surface area and average depth to hold Earth's water. A simple model generates simple formulas that use the volumes of the present continental crust and water for our present Earth to correctly predict the continent crust thickness, ocean water depth, the total continent area, and the total ocean basin area. Using wide ranges of continental crust and water volumes to calculate possible for values early Earth, the crust is invariably more than 24 km thick, the ocean depth is greater than 2200 m, the continent covers more than 25% of the planet, and the ocean covers less than 75%.

5.1 The Distribution and Structure of Continents

We tell in Chap. 3 of the famous hypothesis about the Earth called continental drift. The idea was put first because early maps showed that the shapes of the continents on both sides of the Atlantic Ocean seemed to fit together. Continental drift was strongly promoted by the Meteorologist Alfred Wegener with a series of books describing numerous collections of evidence of the splitting apart and separating between Europe and Africa from North and South.

Figure 5.1 is a modern version of some of his iconic illustrations. It shows how closely they fit together and how a large supercontinent fragmented and spread out. Other evidence was also extremely suggestive. For example, fossil records on both sides of the Atlantic Ocean were similar before about 200 million years ago and dispirit later. The data had long been used to suggest that there were ancient land bridges. Wegener also incorporated geological evidence to suggest that the bridges were either much shorter or absent in the past. His book was widely known, and it

© The Author(s), under exclusive license to Springer Nature Switzerland AG 2024
J. A. Whitehead, *Energy Flow and Earth*, SpringerBriefs in Earth System Sciences,
https://doi.org/10.1007/978-3-031-62694-4_5

included an English Edition. The rates he estimated for some of the drifts proved to be vastly incorrect due to the imperfect dating of the geological record. The mechanism for opening the Atlantic basin was not known. One large objection was that it seemed unrealistic to expect that continents could plow through the Pacific floor because it was simply too strong. Between the early 1920s until plate tectonics was discovered, a few scientists advocated that convection cells in the mantle were responsible, but they were in the minority. Plate tectonics showed that the continents are simply swept around. Ocean floors are created at ridges and injected into the Earth at subduction zones and the dynamic contribution of the continents, if any, was unclear.

Wegener's work continued to suggest that continents have something to do with plate tectonics. Insight into the possible role of continents became stronger after it was recognized in the mid-1960s that there had been sequences of opening and closing of ocean basins. The actual cycle was initially called the Wilson cycle in the mid-1970s after Tuzo Wilson suggested that the Appalachian Mountains had been a location of previous convergence between Europe and North America after a previous version of the Atlantic Ocean basin closed. Then, it was shown that additional cycles of opening and closing of ocean basins occurred much earlier in Earth history and the occurrence and breakup of a few earlier supercontinents were reconstructed. Continents evidently tend to come together every half a billion years or so and form a supercontinent with one giant continent on Earth surrounded by ocean basins. This giant continent breaks into a small number of continents that are then swept by the plates to the other end of the planet to reassemble into another giant continent. Many people use the term supercontinent cycle to describe the formation of previous supercontinents (Pangea, Gondwana, Nuna (or Columbia), Rodinia, and Ur). Their breakups and the use of the term Wilson Cycle referred primarily to the Atlantic (Wilson et al. 2019). The story of continents and how they occupy and dynamically interact with the plates is interesting, and I am working on this set of problems now.

The pronounced difference between ocean basins and continents is a basic fact of our planet. Everyone can see this when looking at a map and globe. The greatest feature is that the surfaces of oceans and continents are at strikingly different elevations compared to the ocean surface with continents at an average 835 m above sea level and ocean floor at an average 3800 m below sea level (Turcotte and Schubert 2002). Figure 5.2a shows the elevation of the continents for present Earth along with the depth of the ocean floor. A well-known double maximum in Earth's elevation versus area plot corresponds to continents and ocean basins. Although small changes to the values are known to occur over time, the great difference between continents and oceans is so obvious that it can be overlooked. The difference is immense and easily quantified. Our ocean basins contain 97% of the world's water but during most of the past few million years when more glaciers were present on the continents, the continental shelves were dry and ocean basins contained almost exactly 100% of the water. Evidently, continents and oceans have come to a balance so that the ocean basins hold almost all of our planet's water. The basins cover approximately 70% of the surface of our Earth to hold all the water. Another interesting fact is that the edge of the ocean water comes almost exactly to the edge of many continental shelf breaks (Wise 1973). Also, the continents are complicated. For example, there are

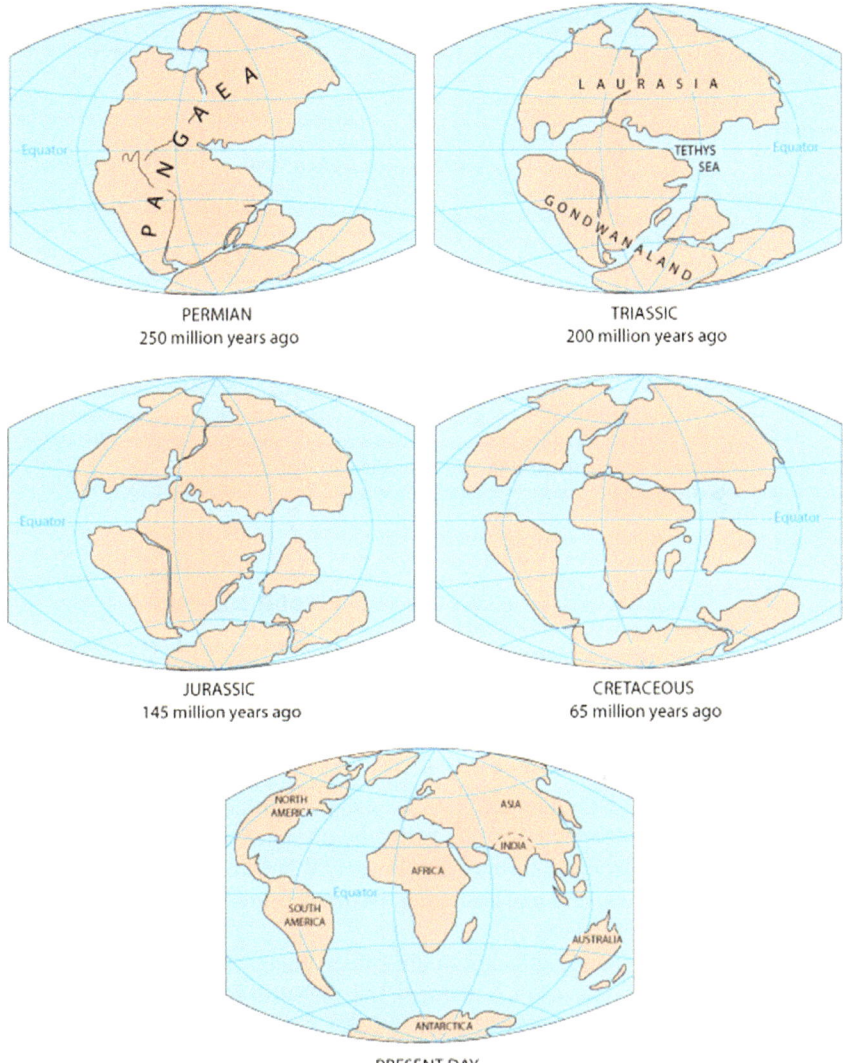

Fig. 5.1 Reconstruction of the movement of continents like that initiated by Wegener (from public domain by USGS)

a few different geological provinces (Fig. 5.2b). How did the continents and ocean basins mechanically evolve so that these two panels occur? How do the oceans have exactly the correct area and average depth to hold Earth's water? What could this possibly have to do with continents and plates? Of course, in this book we also ask what this has to do with the ghost and skeleton of Fig. 1.1?

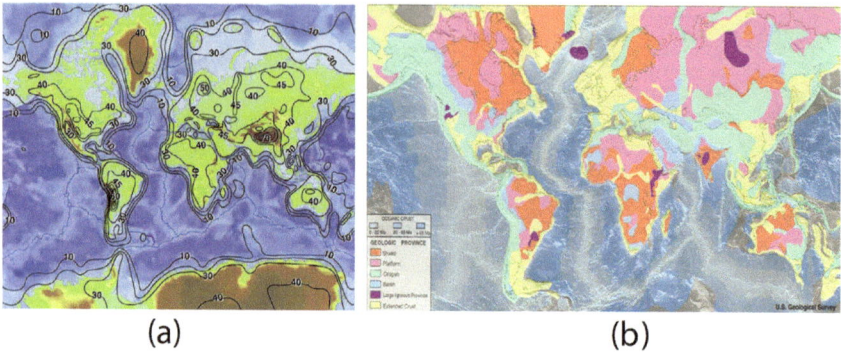

Fig. 5.2 **a** The elevation of land and ocean floor. **b** The different geological provinces in the continents of the world. The shields are cratons (continental material older that a few 100 million years) extending to the surface and the platforms are cratons covered with sediment. The orogens are mountain belts. The large igneous provinces are volcanic areas that have possibly arisen from initiated hotspots (from public domain by USGS—http://quake.wr.usgs.gov/research/structure/CrustalStructure/)

5.2 The Mechanics of Continents and Ocean Basins

It is sensible to ask, "how does continental crust evolve so that the continents and ocean basins of Earth form their present structure?" In the last two chapters, the three greatest components of mantle convection have clearly been identified as subduction zones, hotspots, and upwelling at ridges. In fluid mechanics, we ask "how do components like these determine the size and shape of surface material floating on convection? Do they produce both clustering and a migration of the clusters over time? Do they contribute to the form of the clusters? What are the governing parameters? How big are the clusters and how thick? When do they drift and oscillate back and forth and when not?"

Before examining how continents might affect mantle convection cells, let's review how they are created and maintained. The chemistry of the accumulation of continental crust is actively studied (Taylor and McLennan 2008), but less is known about them as energy-flow structures. Here, we simply assume that the volume of such crust is fixed. The overall dimensions of continents and oceans seem to have been determined by a balance between mountain building that is driven by continent modification by subduction near the edges of plates, which thickens and increases the average depth of continent material, and by erosion, which thins it. Subduction is dynamically important even for collisions between two continent parcels, such as the India-Asian collision that made the Himalayas and the Africa-Europe that made the Alps. Those collisions were driven at first by subduction zones now buried under them. Mountain production at the edge of plates must have been very common on Earth because the ocean floor on each plate is produced at a mid-ocean ridge and is consumed at subduction zones. This means that any fixed thing sitting within a plate, including a continent, migrates to the subduction zones. Therefore, mountain building is a direct consequence of the great rigid "tectonic" plates moving material over Earth's surface. The plate velocities are large enough for the arrival of any continent to a subduction region within a few 100 million years. This time span is brief compared to the approximately 4 billion years that continents have existed (or at least parts of them have). Once a continent arrives at a subduction zone, the continental crust folds and overrides itself again and again from the compressive forces accompanying subduction. The compression produces mountain belts (Fig. 5.3). It is easy to find a list with over 80 mountain building events. The mountain belts are technically called orogens, and they are extensively documented geologically.

Many orogen dates are correlated in time with the supercontinent Cycles. Although each event has a special story, the continental crust is thickened during each event. Peter Clift and I have estimated that in the past 65 my ("recent geological times"), the average continent thickness increased by 2.5% (Whitehead and Clift 2009) mostly in the Himalayas and Andes. This thickening has decreased the total continent area by 1.7% and consequentially increased ocean area.

It is reasonable to assume that in time spans greater than 65 my, the decrease in continent area is balanced to the first order by redistribution of continental material by erosion. Estimates of time scales for the erosion of mountain belts are a few 100 million years, which is much less than the age of Earth. The average surface elevation is presently only 835 m above sea level, so the result of all this is that Earth has tabular continents that lead to the well-known double maximum in elevation

Fig. 5.3 Two sketches of the balance between mantle convection and continent forming material. **a** The effects of different thickening and erosion rates on localized thickening. **b** Localized building of the Andes Mountain chain near the subduction zone in the East South Pacific

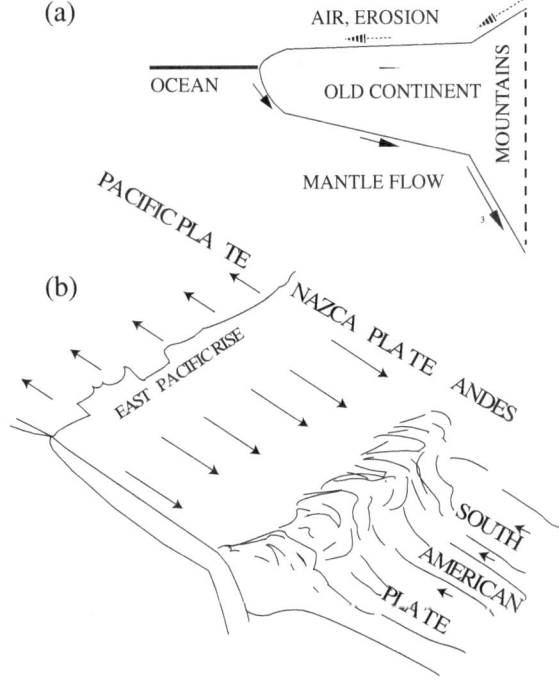

distribution. This double maximum is unique to Earth, and it is not observed for other rocky planets and moons. The eroded continental material deposited on the seafloor is ultimately buried at subduction zones. This feeds volcanism. The cycle of water into and out of the mantle is also involved in generating new continental crust. A clue to the dominance of erosion is that the continent edges are almost exactly at sea level since erosion is more effective under air than under the ocean surface. Figure 5.4 sketches all the factors affecting continents.

An interesting constraint to continent evolution is that the oldest continental regions, called cratons, are generally over 2.5 billion years old. These are still found, which means that they typically have a crustal thickness close to the thickness of younger continental crust. This suggests that the mountain building-erosion balance might have existed over much of Earth's past since the cratons have ages dating back to at least 3.6 billion years ago.

There are various sources that provide new material for continents. The rate of supply from present volcanos is too small to steadily grow continents to the present sizes. Although values are unknown of previous volume flow rates for likely sources of new continent material and of the total volume of free water, it seems safe to estimate that the volumes are within an order of magnitude of the present values for much of Earth's history. The continental material thickness in all the regions away from our present mountain belts is close to 36 km in thickness. With this thickness, the continents, which generally have the density of granite, float in a mantle of denser material. The average continent surface is almost exclusively above sea level, and the average elevation of land above sea level is 835 m (Technically, the form of the continents is called tabular). Tabular continents produce the well-known double maximum in Earth for a plot of the area versus elevation. One maximum of area corresponds to the continent surface and the other to the ocean floor. This double maximum is not observed on other rocky planets and moons.

However, if continent and ocean basin formation is caused by a balance between mountain building and erosion, where does the eroded material go? The pathways of eroded material are complicated. Mountain erosion and biology convert continental

Fig. 5.4 Factors altering the thickness and areas of continents and oceans to first order

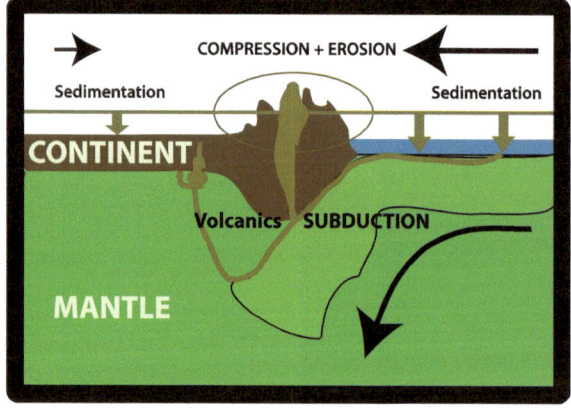

rock to a large variety of sediments. Some sediment remains on the continents, covering the great basins and other older continent regions. Other sediment is swept by rivers to the ocean and deposited on the seafloor where it is swept by plate motions to subduction zones. Since deep ocean trenches are not loaded with millions of years of accumulation of sediment, some of the eroded material is ultimately swept into the mantle. Then, some of this subducted material changes to magma and feeds volcanism because of water and many minerals. Finally, some of the volcanic products contribute to new continent material. Detailed history of sediment distribution over Earth history is certainly a topic of present active research.

A balance exists between mountain building, continent extension, erosion, sedimentation, the burial of sediments in the mantle, the eventual return of some of the sediment material to the continents by volcanism, and the return of water to the surface. Unlike most everything else concerned with mantle plates, if the spreading out of continental material is mostly driven by surface erosion, then the entire process inherently includes the water cycle. How valid is this hypothesized balance? Continent-plate interactions involving continent mechanical formation are in an intermediate stage of development. The development of more data, theory, and calculations continues.

5.3   The Equilibrium Areas and Thicknesses of Continental Crust and Water

The edges of continents are almost exactly at sea level. Wise (1973) suggested that erosion plays a leading role in determining this. This fact can be used for the balance between ocean and thicknesses shown in Fig. 5.5. The principle of isostasy requires that the mantle pressure under the oceans, which is produced by the accumulation of ocean water of thickness d_o and under the mantle of thickness d_m, is close to the same as the pressure under the continents from the continental crust. Here are some simple calculations (Whitehead 2017).

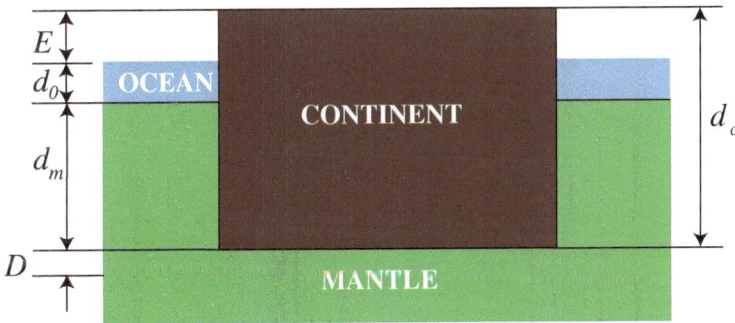

Fig. 5.5 The hydrostatic balance of ocean basins and continents (simplified from Fig. 7 in Whitehead 2017)

Using the following values of the density of ocean seawater $1030 \, kg \, m^{-3}$, continent material $2800 \, kg \, m^{-3}$, and the mantle $3300 \, kg \, m^{-3}$, the hydrostatic balance is $2800 g d_c = 1030 g d_0 + 3300 g d_m + 3300 g D$, here, g *is* the acceleration of gravity. The constant D represents everything left out of this box model, including all the internal structures of the continents and ocean floor, ocean sediment, ocean crust, the oceanic, and continental lithospheres, internal variation within the continents and along their base, ocean ridges, and even the value of E times density of air. These structures clearly have changed over Earth's history but for this first approximation, they are regarded as being forever constant. The largest contribution to D is thought to be layers near the base of continents and cold lithospheres at the base of oceans and continents. These can be crudely estimated to make a value of approximately $D = -2$ km (in this section we will use kilometers for length areas and volumes everywhere rather than meters). The relation $E + d_0 + d_m = d_c$ from Fig. 5.5 allows one to eliminate d_m and the result is that everything depends on the combination of $E - D$.

Will this give a realistic value for d_c? Let us use approximate values for E and D. Starting with $D = 0$ and using the actual value for elevation of continents $E = 0.835$, gives $d_c = 28.1$ km, which is 73% of the presently estimated value of continental crust thickness of 38.4 km (Whitehead and Clift 2009). Next, using crude estimates for the cold bases of continents and the cold lithosphere plus compositional estimates of the deep mantle gives $D = -2$ km. So, using this value along with volumes for present Earth $V_c = 7.679 \times 10^9$ km^3 and $V_o = 1.178 \times 10^9$ (Whitehead and Clift 2009) gives $d_c = 36.53$ km. This is closer to the observed value of $d_c = 38.037$ km (Whitehead and Clift, Table 1). Finally, active erosion rates for the present Earth (Harrison 1998) dictate an approximate value of mean elevation closer to $E = 1$ km. Using a value for continent roots close to that of modern Earth of $D = -2$ km, then the order of magnitude for crustal thickness is 37 km.

Next, we calculate the sensitivity of thicknesses to possible other volumes and areas that might have occured in the past. Values for the mean thickness of the present ocean water $d_0 = 3.8$ km (Whitehead and Clift 2009), continental crust $d_c = 38.037$ km, and the mean continent elevation above the sea surface $E = 0.835$ km (Turcotte and Schubert 2002) gives $D = -2.314$ km. The consequences of different values of the volumes of continental crust and water can be incorporated to find volume V, area A, and the above subscripts for ocean and continent respectively, then $d_c = V_c/A_c$ and $d_0 = V_o/A_o$ where the two areas add up to the surface area of the Earth 5.1×10^8 km^2. If A_o is eliminated, the thickness of continental crust d_c is a function of the volumes of ocean, continent, and present Earth parameters producing the curves in Figs. 5.6, 5.7, and 5.8.

Using the value of the present Earth, a wide range of crustal and ocean water thickness exists. Figure 5.6a shows that continental crust thickness d_c extends from 26 to 60 km thickness. The curves cluster relatively closely together, and this means that continent thickness is relatively insensitive to $E - D$. Therefore, we conclude that continental crust thickness can be smaller than approximately 25 km only if the mean continent elevation is below sea level or the base of continents is radically different from the present one. This result is consistent with the idea that this type of dynamics has existed through much of Earth's history. This idea agrees with the fact

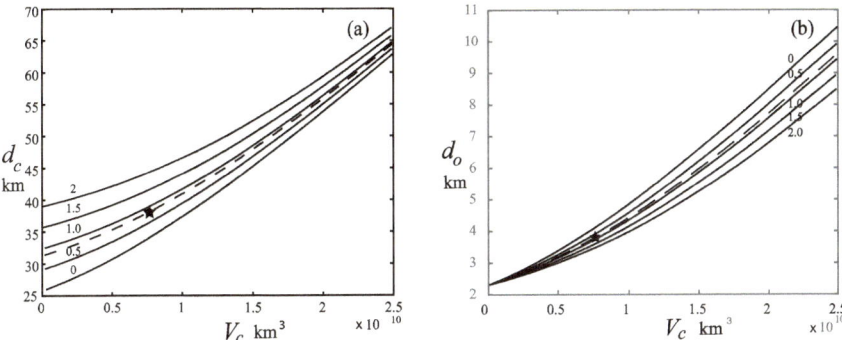

Fig. 5.6 a Mean thickness of continental crust as a function of its volume. The volume of seawater has the present value $V_o = 1.178 \times 10^9$ km^3. The solid curves are for five values of mean continent elevation E above sea level. The dashed curve uses the present value of mean continent elevation. **b** Corresponding mean thickness of the ocean water. The current Earth is shown by a star. Figure 8 in Whitehead (2017)

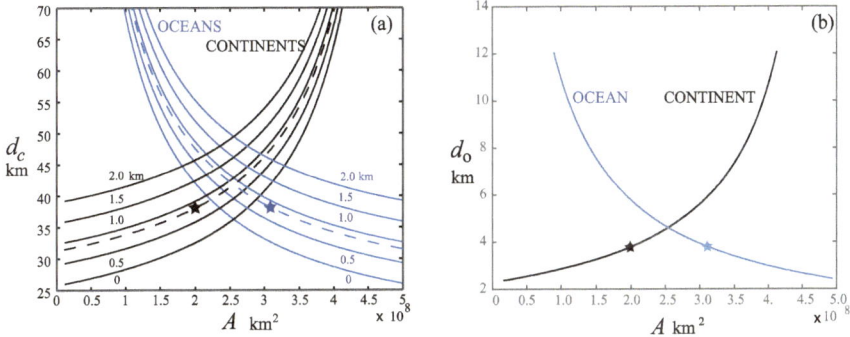

Fig. 5.7 a The thickness of continental crust as a function of the area of continents (black) or the oceans (blue) with all parameters as in Fig. 5.6. **b** The corresponding mean thickness of the ocean water collapses to a single curve. The current Earth is shown by a star. Figure 10 in Whitehead (2017)

that the crust of cratons has thicknesses of the order of present crust (King 2005). The same is true for the depth of ocean water with thicknesses ranging from 2.2 to 10 km.

Crust thickness greater than 25 km is also clear from Fig. 5.7. There is no complete coverage of continents over the Earth with very small ocean basins. Instead, as the volume of the crust becomes immense there is a limit to the area of large continents with the crust thickness close to twice the present value. In contrast, with continents covering about 50% of the Earth or less, the range of possible crustal thicknesses approaches present values (indicated by star). In the limit of small crust volume, the planet is largely covered by water with vanishing continents that have a thickness close to the present values. In summarizing Figs. 5.6 and 5.7, we can say that both

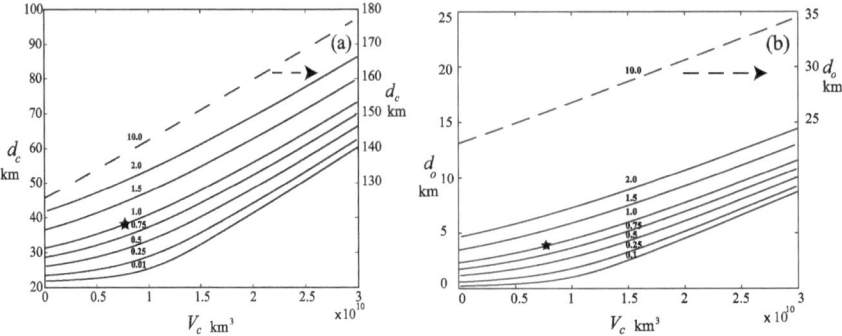

Fig. 5.8 a Continental crust and **b** ocean thicknesses as a function of continent crust volume for eight different values of seawater volume ratio to the present value. Values for the top line are shown on the right axes. The current Earth is shown by a star. Taken from Figs. 11a and 12a in Whitehead (2017)

families of curves show that, given the present volume of water, continent crustal thickness is likely to range between the extremes of 25–40 km and the continent area is less than 75% of Earth's area. Water thickness ranges between 2 and 12 km and it covers more than 255 of Earth's area.

Next, since the volume of ocean water V_o during the past is poorly constrained, various values of V_o/V_c are used to make curves of continent thickness and water depth while varying V_c but keeping the freeboard concept with a mean continent elevation fixed at the present value $E = 0.835$ km (Fig. 5.8). For small values of V_o/V_c that means little water on the planet, the continental crust thickness is greater than about 20 km and the exact value of minimum crust thickness is sensitive to seawater volume. The figure also shows that in general, crust thickness is proportional to V_o. Figures 5.7 and 5.8 together show that for all reasonable parameters, continental crust must be thicker than about 25 km.

5.4 Interaction of Floating Plates and Layers with Convection Cells

Various projects have focused on the driving mechanism of the supercontinent cycles. Elder (1967) developed a theory and a laboratory demonstration showing the possibility that an insulating float of fixed size lying on top of cooling fluid might develop a spontaneous drift. The mechanism of drift is simple. Heating along the top causes fluid to heat up and rise, thereby producing divergence that travels behind and propels the moving float. The interaction of continents and mantle convection was one of my early interests. We sought to explain the Wilson/supercontinent cycle. In the early 1970s, Lou Howard, Willem Malkus, and I analyzed the fluid mechanics of floating surface heaters in a viscous liquid (Howard et al. 1970) and showed that they developed a flow that self-propels the float. Such a float in a laboratory tank traveled back and forth when reflected from the two sides of tank boundaries (Whitehead 1972). I became more excited when numerical studies of two-dimensional convection cells with both insulating and rigid floating blocks showed similar results (Gurnis 1988), as well as when laboratory experiments showed that a thermal insulator floating on top of Rayleigh–Benard convection was also self-propelled (Zhang and Libchaber 2000). Fifteen years later, some numerical calculations showed more details of a self-propelling flow from a floating insulator (Whitehead and Behn 2015). The circulation under the float in Fig. 5.9 is a single cell rather than the pair of cells of ordinary convection cells. We called it "the continental drift convection cell" or "drift cell" for short. This drift cell greatly enhances heat transfer from the mantle to the top. The sinking slab upstream to the drift sinks at an angle rather than vertically downward like a subducting plate sinking into the mantle.

The greatest weakness of floating insulators as a model of continental drift is that the fluid under the insulator is warm, whereas the roots of continents are decidedly colder than the mantle at the same depth. This has recently led to investigations to find another simple model that overcomes this handicap. If the layer of material is only lighter and not thermally active by either insulating or heating, it can also modify convection cells. Recent calculations show that a light layer forms clumps that have cold roots like real continents. Therefore, they seem to be more realistic. Also, in some cases, the clumps will split apart and drift back and forth like the supercontinent cycle. Figure 5.10 shows a side view of a computer-generated flow

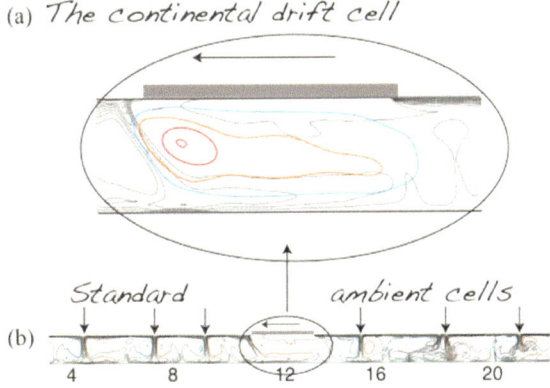

Fig. 5.9 a Side view with gravity pointed downward of the continental drift convection cell produced by a thermally insulating float. Flow lines are shown by color contours and isotherms are in black. The counterclockwise circulation direction under the float propels it toward the left. The tilted cold slab dips under the leading edge. The Rayleigh number is based on internal heating with the value 1.6×10^6. **b** A more distant view of the same drift cell. The convection cells on either side of the moving continent are undisturbed. Red to yellow contours indicate counterclockwise circulation and green, and blue color contours show clockwise circulation (from Whitehead and Behn 2015)

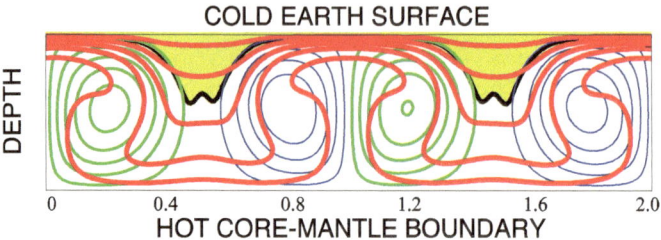

Fig. 5.10 A computer calculation of convection in the presence of a layer of lower density material (yellow with black edge) along the top. Isotherms are red showing a wide cold region under the yellow clumps at the top (from Fig. 6 in Whitehead 2023)

having convection cells with a layer of lower density along the top (Whitehead 2023). The lower density material indicated by yellow is swept by the convergence of the convection cells into clumps of lighter material. These clumps support a cold root like continents.

Figure 5.11 shows a more detailed example with the layer of lower density fluid shown in brown at the top. The ocean basins (blue water drawn in) form between the clumps. The surface elevation of the land and ocean is shown with exaggerated elevation.

The brown fluid representing the continents diffuses down into denser fluid (clear) and collects in clusters along the top because of the convection cells (closed streamlines). These are buoyantly driven by temperature (red isotherms) that is hot at the bottom and cold at the top. Clusters modify the convection cells so that each cold sinking region is split apart. The clusters become shaped like our continents with constant interior thickness and thickening near the edges. The elevated surfaces at the clusters' edges are consistent with mountains built by the cold sinking (subducting) fluid. The clusters and their basements are cold-like continents. Between the clusters, warm denser fluid rises from the bottom and spreads apart at the top. The spreading surface is elevated in the center like mid-ocean ridges.

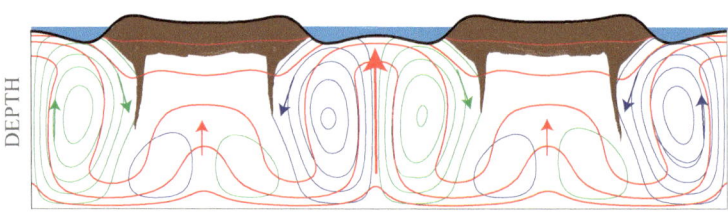

Fig. 5.11 A simple numerical model with a layer of lighter brown fluid lying over convection cells. It gathers in clumps with sinking regions near their edges. Vertical motion under the clumps is small. The region under the clumps is cold because of a lack of upwelling under the clumps and a loss of heat by thermal conduction. The deeper regions have rising hot fluid with the warmest and fastest rising between clumps to generate spreading centers between clumps. Slower rising under the middle of clumps can split them in some cases

In this numerical model, the temperature and density (from different chemicals) diffuse at different rates in the liquid while all other physical properties are constant. This sort of convection is called double diffusive convection. It was discovered at Woods Hole in the 1950s for ocean water, whose density changes with temperature and salinity. Most of the well-known double diffusive flows are either layered or composed of vertical fingers called salt fingers and thus they are very different from Figs. 5.10 and 5.11. I learned about this area of fluid mechanics after arriving at Woods Hole. Our technician, Bob Frazel, had helped Stewart Turner with many of his original experiments (Turner 1985) and he made some of the apparatus. Although scores of applications for double diffusion are now known, this model of the movement and formation of continents and ocean basins is a new application.

If the fluid is internally heated, or if the flow is slowly changing as in this figure, a warm plume slowly forms deep below each cluster and rises, splitting each cluster apart. Clusters drift back and forth laterally and continue to cyclically form and re-form like the Wilson/supercontinent cycle. Figure 5.12 shows a flow with such a periodic cycle. In this example, all the heat flow comes from internal heating within the fluid. The start and end of one-half of a complete cycle are shown in the two panels. Each cluster splits, and each half drifts apart only to merge with a neighbor, making a new blob of the lighter component. This half is followed by a return to the starting arrangement that completes the cycle.

Most other models for continent/convection interaction are more complex (Bobrov et al. 2022). Some have flow in a sphere and convection's effects on Earth's heat budget when cycles are included. Some results show that deep mantle heating relates to deep plume formation. Many models have values of density for continental and ocean crust, large viscosity at the surface, lower at middle depths, and high viscosity of the deep mantle. Many studies have mechanically strong insulating plates over

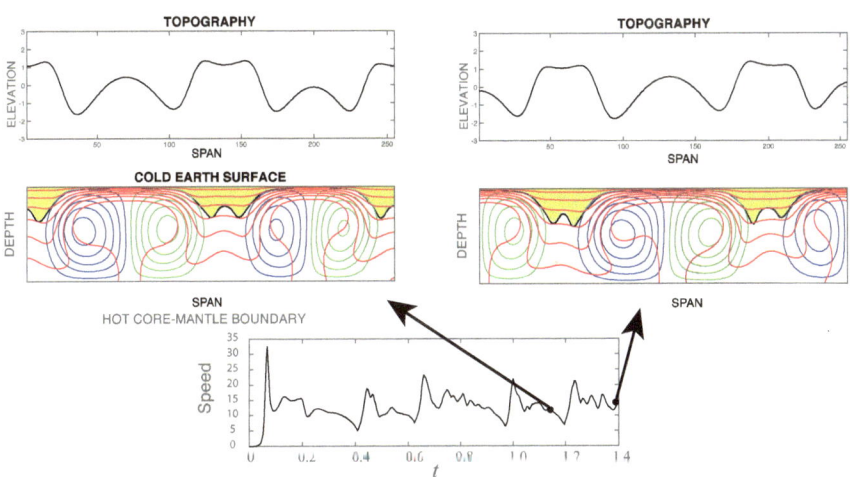

Fig. 5.12 Flows at opposite times of the cycle shown at the indicated times

mantle convection. None of them include erosion or the water cycle. The inclusion of thermochemical effects into numerical models has advanced greatly, too. These produce a lower mantle influence on the production of plume cycles, effects from the core, and continent-like clusters near the surface. Agreement with results from seismology, bulk composition data, elevation reconstruction, and polar wander has steadily improved. High resolution models presently extend back approximately a billion years and partially duplicate the Wilson/supercontinent cycle, but this is only about 30% of Earth's age. In complicated numerical models it is difficult to separate which components are needed for predictive skill and which are simply adjusted to produce a good agreement with the present Earth. It is to be hoped that ultimately, the overall picture that we sketched at the beginning of this chapter of the supercontinent cycle will be confirmed. In summary, convection in the mantle conveys the continents to many of the sinking regions. Thickening of continent material occurs from mountain building and volcanism. However, so far thinning is not by erosion.

Numerical modelers face a great number and variety of challenges. The entire evolution of all the plates, along with the continents within them, is challenging because of the complexity of the flows along with the variations of strength and viscosity that are expected for Earth-like materials over such wide ranges of temperature and pressure. A major challenge comes from the transition to a solid near the surface of Earth. Although plate reconstructions have been made for the past tens of million years, uncertainly remains about the evolution of the margins of the plates in general and especially about the continent margins over hundreds of millions of years. Peter Clift and I have estimated that during the past 65 million years the continent area has been decreased by about 1.7% as the present Earth became more mountainous as the Himalayas, Andes, and Alps formed. Since no sizeable sources of new continent material are known over that period, the conserved volume of continental crust means that an area decrease is accompanied by a net increase in average thickness. We estimate that the average continent elevation relative to the average ocean floor has increased by 54 m and average ocean water thickness has decreased by 102 m resulting in continents rising 156 m over sea level over that period. This rise is partly balanced by continent erosion, largely within the same mountain belts that caused the shortening.

The influence of continents on mantle convection is becoming clearer, too. The basements of continents on Earth are cold, simply because they are old and have been cooled by being close to Earth's surface for many hundreds of million or even many billion years. Many models with internal heating of the mantle generate hot regions under continents. If the model has heating in the deep mantle, it occurs more from the absence of subduction induced by continents than from insulation from continent material.

5.5 Summary of Energy-Flow Rates for Continents and Ocean Basins

The overall energy-flow rates associated with both continent thickening and thinning as well as for the Wilson cycle are quite complicated. Mantle convection is driven by the flow of energy from internal heat generation within the Earth. Meanwhile, erosion involves the global weather cycle driven by energy flow from the sun. Figure 5.13 is a sketch with some approximated estimates of the values of energy flow in this cycle. Some values are from calculations done in the previous chapter and some values are calculated in Chap. 6.

For mantle convection, heat flow enters the bottom of the mantle at the rate of 2.3×10^{12} W and the internal generation of heat within the mantle might be 2×10^{13} W. If the mantle is only very slowly cooling at the rate of 3.7×10^{12} W, then there is almost a perfect balance between heat production within the mantle and the cooling of the plates of 2.6×10^{13} W (see Table 4.1 of the previous chapter). This flow rate of thermal energy is converted to mechanical energy to propel plate motion, and a small amount is used to laterally compress the continent material and build mountains near the margins.

The energy cycle for erosion is independent of the mantle cycle. The originating energy source for erosion is the water cycle driven by solar heating that has the much greater rate of 5.5×10^{16} W onto the entire planet. Solar heating produces the

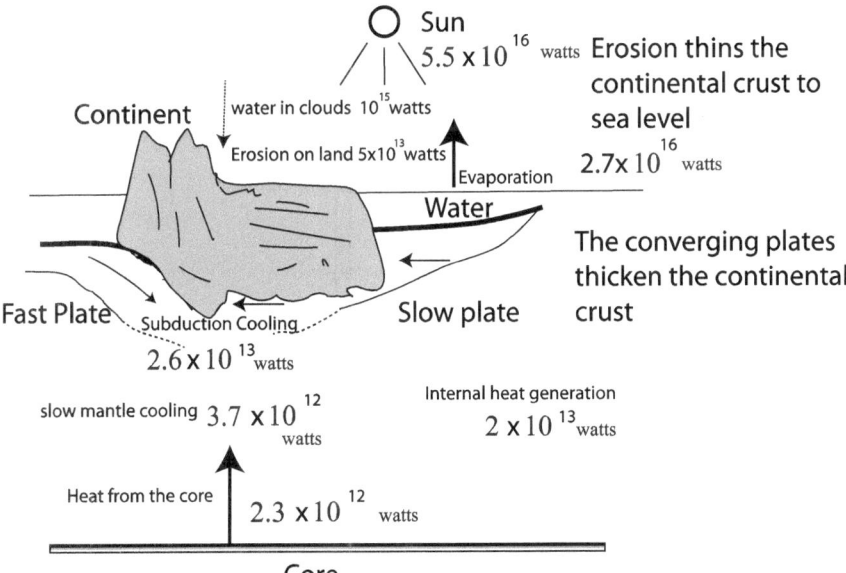

Fig. 5.13 Energy balances of continent and ocean basin formation. These energy sources and flows have produced our present Earth with continents occupying approximately 1/3 of the surface and the ocean basins 2/3 and with almost all water in the ocean basins

evaporation of water at a rate of 2.7×10^{16} W (the magnitudes shown in Fig. 5.8 are calculated in Chap. 6). The evaporated water is elevated to the clouds, and the part that falls as rain on the land erodes the continental material. Potential energy is liberated as water rains from the clouds. It flows downhill and erodes the land. The rate of potential energy release is calculated by multiplying gravity times the volume flow rate of the water times density times elevation. This simple estimate was one of the first calculations that I did when arriving at Woods Hole in 1971. Take the rate of potential energy change for evaporated water moving from the ocean surface to 10 km elevation where clouds begin to form raindrops. We define this formula for the rate of release or gain of potential for a flowing liquid in Fig. 2.12. It is

$$Er = g \times \Delta\rho \times d \times Q.$$

The values in this calculation are the strength of gravity $g = 10$ m/s^2, the density difference between air and water $\Delta\rho = 1030$ kg/m^3, the elevation of the water to 10 km, or $d = 10^4$, and a value of Q that is estimated from the fact that the oceans evaporate approximately 1 m of water each year. Because the size of Q occurs over the ocean area of 3.4×10^{14} m^2, this value times one meter divided by the number of seconds in a year comes to approximately $Q = 10^7$ m^3/s. Using these values, the rate of potential energy gain of water that evaporates and rises to cloud elevation to form precipitation is $Er = 10^{15}$ W. This value is not trivial. It is 1/55 of the total input of energy from the sun. The potential energy that is available for flowing water to erode the land by mechanical work is considerably smaller for two reasons. First, all the water that evaporates does not fall as rain on land, and second, the mean elevation of the continents is 860 m. An estimate of potential energy released by water flowing downhill might be as small as $Er/20 = 5 \times 10^{13}$ W. This value is higher than the internal heat production of the mantle of 2×10^{13} W. A better calculation must include the energy involved in the weathering of rock and energy losses in turbulent water that is not breaking the land down and sweeping it laterally. This present calculation is very crude and could be greatly improved, but it indicates that there is plenty of energy available for the balance between mountain building and erosion. To compare magnitudes, human energy consumption is 2×10^{13} W.

References

Bobrov A, Baranov A, Tenzer R (2022) Evolution of stress fields during the supercontinent cycle. Geodesy Geodyn 13(4):363–375

Elder J (1967) Convective self-propulsion of continents. Nature 214(5089):6–750. https://doi.org/10.1038/214657a

Gurnis M (1988) Large-scale mantle convection and the aggregation and dispersal of supercontinents. Nature 332(6166):695–699. https://doi.org/10.1038/332695a0

Harrison CGA (1998) The hypsography of the ocean floor. Phys Chem Earth 23:761–774

Howard LN, Malkus WVR, Whitehead JA (1970) Self-convection of floating heat sources: a model for continental drift. Geophys Astrophys Fluid Dyn 1(1–2):126–142

King SD (2005) Archean cratons and mantle dynamics. Earth Planet Sci Lett 234(1–2):1–14

Taylor S, McLennan S (2008) Planetary crusts: their composition, origin and evolution (Cambridge planetary science). Cambridge University Press, Cambridge. https://doi.org/10.1017/CBO978 0511575358

Turcotte DL, Schubert G (2002) Geodynamics, 2nd edn. Cambridge University Press

Turner JS (1985) Multicomponent convection. Annu Rev Fluid Mech 17:11–44

Whitehead JA (1972) Moving heaters as a model of continental drift. Phys Earth Planet Inter 5:199–212

Whitehead JA (2017) Dimensions of continents and oceans—water has carved a perfect cistern. Earth Planet Sci Lett 467:18–29

Whitehead JA (2023) Convection cells with accumulating crust: models of continent and mantle evolution. J Geophys Res: Solid Earth 128. https://doi.org/10.1029/2022JB025643

Whitehead JA, Behn M (2015) The continental drift convection cell. Geophys Res Lett 42:4301–4308. https://doi.org/10.1002/2015GL06448

Whitehead JA, Clift PD (2009) Continent elevation, mountains, and erosion: freeboard implications. J Geophys Res: Solid Earth 114:B05410. https://doi.org/10.1029/2008JB006176

Wilson RW, Houseman GA, Buiter SJH, McCaffrey KJW, Doré AG (2019) Fifty years of the Wilson cycle concept in plate tectonics: an overview. Geological Society, London, Special Publications. https://doi.org/10.1144/sp470-2019-58

Wise DU (1973) Freeboard of continents through time. Mem Geol Soc Am 132:87–100

Zhang J, Libchaber A (2000) Periodic boundary motion in thermal turbulence. Phys Rev Lett 84(19):4361–4364. https://doi.org/10.1103/PhysRevLett.84.4361

Chapter 6
Energy Flow and Producing the Earth's Magnetic Field—The Dynamo

Abstract A historical review of mechanical dynamos is given. It shows how mankind learned how to make an electric field and a magnetic field sustain each other to produce electricity. Like convection, a value of forcing must be exceeded before the production of an electromagnetic field can exist; hence the electromagnetic field is an energy-flow structure. Flow in Earth's metallic liquid core was known to be one of the most likely candidates to produce Earth's magnetic field, but calculation of the fluid flow needed for this production was not successful until the 1970s. Solving how this flow works (called the dynamo problem) required the development of electromagnetic fluid dynamics studies. Finally, full dynamo calculations were completed on numerical computers so that today many of the features of Earth's dynamo are being investigated.

6.1 Mechanical Dynamos

Rayleigh's calculation, outlined in Chap. 2 shows how a cellular structure grows from background noise. The motionless fluid was in a field of gravity in a fluid layer with warm temperature at the bottom boundary and cold at the top. Growth occurs only when a certain condition exists where a collection of parameters exceeds a fixed value. That collection is now universally called the Rayleigh number. Generally, its value must exceed roughly Ra = 1000 for convection cells to form. The Rayleigh number for the mantle is estimated to be between 10^6 and 10^9, so that the cellular nature of the tectonic plates makes sense. Here, I will describe how the magnetic field of the Earth is also formed by instabilities, and that a critical value of a magnetic version of the Reynolds number is required (Reynolds number expresses the ratio of flowing fluid inertia to viscous forces. The magnetic version is the ratio of inertia to electromagnetic forces).

Magnetism itself was a mystery in ancient times and somewhat puzzling. Magnetic material was known and finally developed to the point of being useful for compasses that were incredibly useful for traveling. In the early 1800s, as the science of electric fields advanced, experiments showed that magnetic forces accompany electric currents. Michael Faraday found that moving a wire near a magnet generated an electric current and that an electric current in a magnetic field produces a force on the wire. His experiments were fundamental to the invention of numerous devices that were discovered during the industrial revolution such as electric dynamos, electric motors, telegraphs, telephones, and electric lights.

It was a staggering development to produce electric and magnetic fields mechanically with electric dynamos. One of the simplest generators of an electric current was made by Faraday in 1831–2, who used a metallic disk that rotates between the two poles of a permanent magnet. A brush along the outside of the disk acquires a different voltage from the shaft at the disk center, and a closed electric circuit between the brush and shaft develops a small current. Although Faraday gets credit for these experiments, other independent experiments were conducted in 1827 by the Hungarian Priest, Anyos Jedlik. The first generator was inefficient, but subsequent devices used multiple windings of wire that worked much better. The properties of magnetism were understood well enough at that time so it became clear that the permanent magnet could be replaced by an electromagnet as sketched in Fig. 6.1. In this case, the magnetic field was thought (at first) to need a supplemental field to be started and produce the self-sustaining current and magnetic field. It soon became clear that the electric current could grow spontaneously if the machine spun rapidly enough. This electric dynamo was an invention that revolutionized the world.

Basically, Fig. 6.1 is a sketch of a simple configuration of wires and a disc that rotates so that an electric current and a magnetic field both spring forth out of nothing. The wire coil replaces the permanent magnet and makes a self-sustaining dynamo. It produces a magnetic field from an electric current J with magnetic field lines in the figure shown as B. This device was called a homopolar dynamo and was the first dynamo in existence to replace the dynamo with a permanent magnet. It is self-sustaining and must exceed a set speed before it a tiny instability to the electromagnetic field grows. Therefore, the electromagnetic field and the electric currents are just like the convection cells in Chap. 2 that grow from a tiny perturbation. This particular dynamo also has received recent interest because it exhibits chaotic behavior when driven by a constant torque rather than constant speed. Chaos theory is covered in many other books and won't be described here. From about 1850 on, this device has been extended in numerous ways for engineering purposes.

Fig. 6.1 Sketch of the disc dynamo where a rotating disc sweeps through the magnetic field B to make a current. The current is fed through slip rings to a wire above the disk where it makes field B

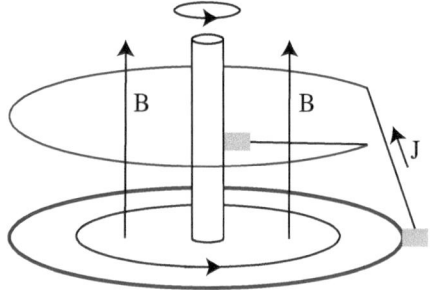

Let's clarify some aspects of the energy flow within this device. First, let's start with the small electric current in the coil. It produces a magnetic field with the magnetism directed upward as shown in Fig. 6.1. This field is indicated by the straight arrows with the strength marked B. The energy stored in this field is proportional to B^2. Second, this magnetic field interacts with the spinning disc to pump electrons toward the center and makes the current that flows in the direction needed to strengthen the field. This flow of electrons also contains energy. (Note that the convention we use here for the current direction is the same as the one used in physics: The current J flows in the direction of the positive charge, which is opposite to the direction of the flow of electrons.) Third, as with cellular convection, with an imposed spin rate of the disc, there is a collection of parameters whose product must be exceeded to make the small current grow. Fourth, the flow of electrons in the magnetic field produces a force that resists the rotation of the disk. Therefore, driving the disc requires mechanical work to make the energy. For steady rotation and a fully developed dynamo, the energy of work driving the dynamo is dissipated by the flow of electrons in the conducting metal, which produces heat. In addition, of course there is the friction of the disc bearings and brushes. Therefore, this electric field and current along with the magnetic field are close to being energy-flow structures except that the structure of the wires and disc are imposed. Energy flow is needed to produce them, and they have losses in energy from electrical resistance and friction. One important detail is that the device depends on the direction of rotation of the disc compared to the winding of the coil. If the disc rotates in the wrong direction, this dynamo configuration does not work. This gives us a hint that making a dynamo using only the flow of a conducting fluid that could end up with a full energy-flow structure might not be simple to do.

Ever since the compass was discovered, people had wondered what made the needle point in one direction. Other materials that were given the general name of lodestone would deflect compass needles, so it was thought that possibly giant lodestone deposits might be responsible. Numerous questions about Earth's magnetic field existed well into the nineteenth century. The compass worked everywhere, so was there some reason that the hypothesized lodestone lined up the Earth's rotation axis? Did the North Star play a role? After Faraday's experiments, a connection between electric currents and magnetism was known, and people wondered whether natural movements such as ocean currents or even winds produced the Earth's magnetic field.

By approximately 1880 the theory of electromagnetism had been well developed. A clear linkage between electricity and magnetism was expressed by four calculus equations known as Maxwell's equations. They provided a good and complete expression of electromagnetism in a medium that is not moving. A fundamental finding was that the equations express electromagnetic waves, and this paved the way to the development of studies of electromagnetic waves over a wide range of frequencies including light waves, radio waves, microwaves, X-rays, and many others. Scientific and practical devices were developed to study and use electromagnetism and light. It didn't take long for scientists to discover that a magnetic field altered the light from glowing material. The light emitted by sunspots, which had been observed for centuries, indicated strong magnetic fields, and the light from other stars indicated that there were magnetic fields in them, too. This was puzzling because when ordinary lodestone magnets were heated up to a high enough temperature (the Curie temperature), they lost their magnetic properties. Therefore, it seemed likely that electromagnetic fields in stars are not made by ordinary magnets, because stars are much hotter than the Curie temperature. Moreover, heat was measured to be flowing out from deep in the Earth, and most calculations indicated that the temperatures near the center of the Earth are likely to be much higher than the Curie temperature. Therefore, until about 1950, people were uncertain about how the Earth's magnetic field is produced.

One suggestion was that an electrically conducting fluid (such as iron in the core of the Earth) flows so that a magnetic field grows out of nothing. A thorough review is given a thesis by Mason (2023), which can be downloaded online. The process is self-sustaining so long as there is an energy source sufficient to maintain convection. The calculus equations were known, but solutions to them were not easy to find. For one thing, it was shown early in the twentieth century that the flow cannot be strictly two-dimensional. The flow needs to have a corkscrew swirl. An important step toward the development of dynamo theory for a flowing liquid was to find exact flow patterns that make a self-generating magnetic and electric current field. These flows were called kinematic dynamos, because the flow pattern is prescribed (just like we prescribed the wires and currents in Fig. 6.1). The fluid mechanics needed to produce the flow patterns was ignored. Various kinematic dynamos already existed when I was introduced to the problem. One is called Herzenberg's dynamo, made of two cylinders with their spinning axes at right angles to each other inside of an electrical conducting material. A laboratory demonstration by Lowes and Wilkinson (1963) had two rotating cylinders of an iron alloy embedded in a solid block of the same material and used liquid mercury for good electrical contact. High rates of spinning produced an electromagnetic field. Some kinematic dynamos are purely mechanical and close to the fluid flow equivalent to Fig. 6.1, and others are quite elaborate. Later, there were some statistical models of turbulent fluid flow with overall features leading to a magnetic field.

6.2 Experiments and Theory

When I was a graduate student, no liquid dynamo had ever been made. When I went to UCLA as a postdoctoral investigator in 1968, Willem Malkus was trying to produce a spontaneous electric current and magnetic field using a spinning spherical cavity filled with liquid sodium that was precessing. A precessing body has a spinning axis that is also rotating. Before doing the experiment with sodium, he used a model of the cavity with a transparent sphere filled with water to determine the flow pattern and to estimate how rapidly to spin and rotate (precess) the axis for a likely dynamo. The hope was that the flow might have enough three-dimensional complexity to make a dynamo. Positive results would suggest that the precession of the Earth might produce enough swirling in the liquid core to make Earth's magnetic field. When he used liquid sodium in a larger experiment to seek any signs of magnetic field generation, the closest he got was to measure the decay time of waves that were produced by an externally excited magnetic field (unpublished, private communication). The waves decayed and indicated that the experiment was getting close, perhaps a factor of about 2 too small. I listened in awe as visitors came and gave enthusiastic support. I cherish fond memories of numerous exciting seminars on progress in the dynamo problem.

The idea put forth vigorously by Malkus (1968) was that precessing the spinning liquid metal in the core might make Earth's geodynamo. It still has some support, but a more likely candidate in the Earth and some planets is the thermal convection cells within a rotating liquid core. The ideas about the dynamo problem for Earth and planets were developed in hundreds of papers from about 1915 onward.

Kinematic dynamos were well developed by the time I came along. Dynamic dynamos are a result of conditions at the boundaries of the fluid. One cannot impose a magnetic field and the flow field would evolve along with the electromagnetic field. It was expected by all the experts that one could start with boundary conditions that would produce a known flow field, and then when a dimensionless number called the magnetic Reynolds number was above a certain value, a dynamo might occur. Finding forces imposed at the boundaries needed to achieve a sufficient speed, and the flows that twisted any perturbations enough to make an electromagnetic field was difficult. In addition, the possibility that turbulence might come forth before the electromagnetic forces grew made progress difficult.

The fluid mechanics of the Earth's dynamo from the convection of conducting fluid in rotating spheres was developed. The rough ideas were sketched out by Bullard (1949), but more refined calculations proved to be difficult to develop. It was learned that the rotation of the Earth tends to make convection cells in the liquid core spiral around axes that are aligned with Earth rotation. The size of the cells in a north–south direction is greatly modified as these cells move toward or away from the rotation axis. Complexity also arises as the cells interact with the curved surfaces of the inner core and the core-mantle boundary. All these aspects were increasingly included in fluid mechanical studies in the 1970s and 80s, although people were not sure exactly what to include. There are effects from inner core wobble and other variations in Earth rotation from the moon that might supply some energy of the dynamo. Despite the challenge of including all these factors, numerical simulations of both the Earth and of other planetary cores and their magnetic fields have became highly developed.

I watched numerous struggles by gifted theoreticians as they gradually began to find the keys needed. Flows that are rapid enough to make a dynamo would generally be turbulent, and this barrier seemed almost impossible to overcome. In the process, something became known about the statistics that a turbulent field would need to produce a magnetic field. Figure 6.2a shows magnetic fields calculated to grow in convection cells in a rotating spherical shell of conducting fluid. Cuong and Busse (1981) calculated a linear stability prediction that is only a starting estimate of the dynamo that accompanies the calculated convection cells, although it is a very useful one since it predicts the conditions for the growth of an instability (like Rayleigh's stability calculation in Chap. 2). Of course, it does not calculate the fully developed flow. This pattern might be expected to grow until it is limited by either viscous drag or the electromagnetic forces from electric currents flowing within the magnetic field. Flows limited by viscous drag were the easiest to find, but their speed was so slow that the magnetic field had energy values much smaller than the kinetic energy of the flow itself. Estimates of the Earth's field indicated that growth is probably limited by the electric currents rather than viscosity. The electric currents produce heat just like a household electric heater. Estimates of the drag can be related to the expected decay time of a current that would exist in a motionless material with the same electrical properties of the core. This decay time is estimated to be a few 100 years based upon the properties expected for liquid metal at the core's temperature and pressure.

It has become generally accepted that the energy source for Earth's magnetic field is thermal convection with cells that are modified by both Earth rotation and the inner core. The cells were originally thought to be driven by heat flow from the core up into the lower mantle, but as properties of the inner core became more clearly known, it was realized that convection cells are driven by buoyancy forces from variations in liquid composition rather than the coefficient of thermal expansion like the cells in Chap. 2. Liquid composition variation occurs from the accumulation of the inner

Fig. 6.2 The axisymmetric magnetic fields in a growing perturbation leading to a dynamo in a spherical shell. The fields have two different components. The top panels show the poloidal field, which in this example is like the common north–south permanent magnet, but in general it has higher harmonics in the core. The poloidal field is accompanied by an electric current circulating around the spherical shell. The bottom panels show the second field that is called the toroidal field. It only exists within the conducting fluid. Figure 1 in Cuong and Busse (1981) used under Creative Commons CC-BY license

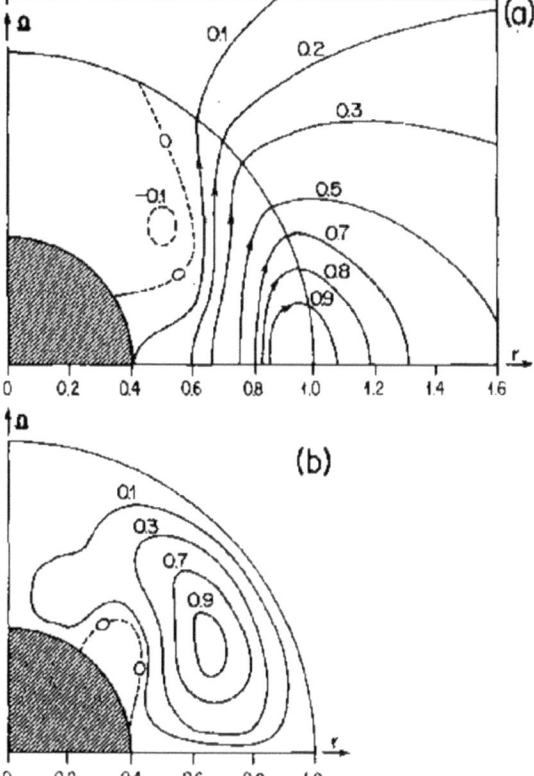

core material because it has a slightly different composition from the liquid core. The inner core is slowly growing as the Earth cools. In the process, the different density of the liquid from core formation probably drives the motion. Otherwise, the convection cells are the same as the flows in numerical studies and in the laboratory for ordinary liquids. In addition, there might be heat production from radioactive decay in the deep Earth. Possible locations are poorly known.

Although Fig. 6.2 is quite clear, it proved to be very difficult to calculate flows and fields when they grow to a finite size. Quite possibly, the flow becomes so rapid that is collapses to a very disordered flow. During this collapse, many harmonics of the field grow in addition to the ones in Fig. 6.2, and their strengths change continually in every case calculated so far. Actual detailed mathematical solutions have been slow in coming.

Two large projects finally produced actual flowing liquid dynamos. They both significantly helped us to understand both the dynamics of fluid dynamos in general and Earth's dynamo in particular. First, a breakthrough came from numerical studies with supercomputers, where one of the barriers was removed by neglecting fluid inertia. Also, some of the issues of the conditions of magnetic fields at the boundaries led to simplifications. Even so, the calculations were massive and only completed after years of development. Results were reported in 1995 by two groups, one in Japan and the other in the United States. Both projects reported the appearance of magnetic fields for thermal convection in electrically conducting fluid in a rotating sphere. The total energy in the magnetic field was many times larger than the kinetic energy of the flow. The irregular motion within the numerical program is extremely complicated. The flow fields of velocity, electric current, and magnetic field distributions are unsteady and hard to understand. Figure 6.3 shows three different times in the evolution of the field for the US model. At the beginning, there is a large North–South (dipole) component to the magnetic lines like the one on Earth right now. At two other times roughly 500 years apart, a transition from this dipole to a dipole of the other sign (a "north to south pole reversal") is occurring. During reversal, the large and familiar North–South dipole field that we are all familiar with becomes small, and the region is filled with smaller scale magnetic lines before the opposite sign grows. Earth's field has such reversals, and this feature leads to the time dependence that helped the discovery of plates spreading from mid-ocean ridges. The production of reversals in the numerical model was a welcome surprise.

The numerical projects invariably have complicated convection cells so that the fields are very highly structured and change with time. The familiar north–south magnetic structure is only one component of the much more complicated field. Time variations are always present and there is some hope that the smallest and fastest fluctuations on Earth might be detected by satellite measurements soon to compare with the numerical results.

The second large set of projects was reported roughly 5 years after the successful numerical calculations of the magnetic fields were published. In a giant laboratory experiment, magnetic fields were produced using pumped liquid metals (called magnetohydrodynamic experiments). The technology to pump liquid metals with externally imposed magnetic fields was well developed because liquid metals needed

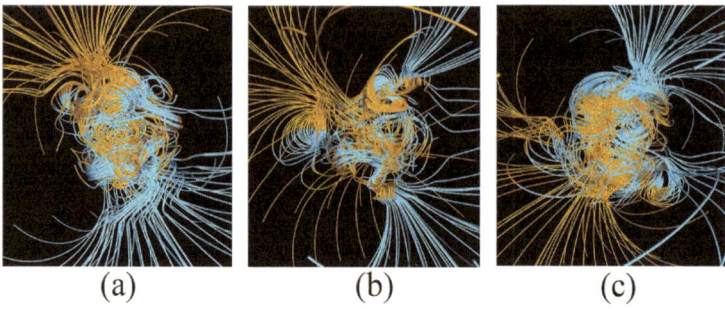

(a)   (b)   (c)

Fig. 6.3 Magnetic lines in numerical calculations for a liquid core with spinning convection cells as a reversal occurs over an interval of 1000 years. North away from Earth is yellow and South toward Earth is blue. The rotation axis is vertical, so the dipole is tilted (Images supplied by Gary Glatzmaier, like Fig. 1 in Glatzmaier and Roberts 1995)

to be pumped to cool nuclear reactors. Although the technology existed, having a flow spontaneously produce such a field was undocumented. Individual scientists in the U. S. could not raise the money needed to fund the large projects that were required, and other countries organized the research more effectively. One experiment in Latvia (Gailitis et al. 2000) had spiraling flow down a long tube that was surrounded by an annular region with backflow, both being immersed in liquid sodium. The first confirmation signals were a growing oscillation. Another experiment in Germany (Stieglitz and Müller 2001) produced magnetic fields in flowing liquid sodium through an array of guide-tubes with opposite and counterrotating flows. In both cases, the flows became intense enough to twist magnetic lines sufficiently for self-amplification. The costs and engineering of the projects verified the conjecture that the support be supplied on a national scale. An experiment in France (Monchaux et al. 2007) had flow driven by two counterrotating disks in a cylindrical cavity filled with liquid sodium. The disks were facing each other and had impellor blades to produce shear layers and intense turbulence. Stainless steel disks did not work but iron disks with higher magnetic permeability worked. Results have demonstrated many of the features expected by theory (Monchaux et al. 2009), such as a critical value to start, transitions to more complex flows, and oscillations. The power required was large, and the sodium had to be cooled to avoid overheating.

For Earth, the magnitude of the flow of heat from the core to the lower mantle might be as large as 2.3×10^{12} W. This value is close to our estimated hotspot heat flow. Estimates of the mechanical efficiency of convection cells and the electromechanical efficiency of dynamo production are difficult to produce so far, because the numerical simulations don't use material properties that are verified in detail. The actual dissipation on the electromagnetic field by electrical conduction is not resolved yet because of small-scale turbulence. Possibly the magnitudes of the energy flow in the dynamos are smaller than the hotspot values but within the same order of magnitude. At least, this seems to be the case if one uses an estimate of the power ranging from 0.6 to 1.9×10^{12} W (Roberts and Glatzmaier 2000). Estimates of the thermal energy associated with inner core formation are very uncertain but can range from the hotspot value to possibly a few times greater than this.

This ends the story of this remarkable development. I was not directly involved, but I watched the progress and know most of the theoriticians. Following the intricate fluid mechanics was challenging for me. It was helpful that our present understanding of rotating fluid mechanics applies to flow in the Earth's core, just like it applies to oceans, atmospheres, planets, and some stars. Review articles by Busse (2000), and Glatzmaier and Olson (2005) helped me to decide to write this chapter because their history is part of my own recollections. In addition, much of the progress was discussed in the Geophysical Fluid Dynamics Summer Program at the Woods Hole Oceanographic Institution to my delight. Otherwise, an isolated fluid dynamicist turned oceanographer and Earth scientist like me would not have been able to write this summary!

References

Bullard EC (1949) The magnetic field within the earth. Proc R Soc Lond Ser A Math Phys Sci 197(1051):433–453

Busse FH (2000) Homogeneous dynamos in planetary cores and in the laboratory. Annu Rev Fluid Mech 32(1):383–408

Cuong PG, Busse FH (1981) Generation of magnetic fields by convection in a rotating sphere, I. Phys Earth Planet Inter 24(4):272–283

Gailitis A, Lielausis O, Dement'ev S, Platacis E, Cifersons A, Gerbeth G, Gundrum T, Stefani F, Christen M, Hanel H, Will G (2000) Detection of a flow induced magnetic field eigenmode in the Riga dynamo facility. Phys Rev Lett 84(19):4365–4368

Glatzmaier GA, Olson P (2005) Probing the geodynamo. Sci Am 292(4):50–57

Glatzmaier GA, Roberts PH (1995) A three-dimensional self-consistent computer simulation of a geomagnetic field reversal. Nature 377(6546):203–209

Lowes FJ, Wilkinson I (1963) Geomagnetic dynamo: a laboratory model. Nature 198(4886):1158–1160

Malkus WVR (1968) Precession of the earth as the cause of geomagnetism: experiments lend support to the proposal that precessional torques drive the earth's dynamo. Science 160(3825):259–264

Mason SJ (2023) The strong-field regime of the geodynamo. Doctoral dissertation, Newcastle University

Monchaux R, Berhanu M, Bourgoin M, Moulin M, Odier P, Pinton JF, Volk R, Fauve S, Mordant N, Petrelis F, Chiffaudel A, Daviaud F, Dubrulle B, Gasquet C, Marie L, Ravelet F (2007) Generation of a magnetic field by dynamo action in a turbulent flow of liquid sodium. Phys Rev Lett 98(4):044502

Monchaux R, Berhanu M, Aumaître S, Chiffaudel A, Daviaud F, Dubrulle B, Ravelet F, Fauve S, Mordant N, Petrelis F, Bourgoin M, Odier P, Pinton J-F, Plihon N, Volk R (2009) The von Kármán sodium experiment: turbulent dynamical dynamos. Phys Fluids 21(3):035108. https://doi.org/10.1063/1.3085724.hal-00492266

Roberts PH, Glatzmaier GA (2000) Geodynamo theory and simulations. Rev Mod Phys 72:1081–1123

Stieglitz R, Müller U (2001) Experimental demonstration of a homogeneous two-scale dynamo. Phys Fluids 13(3):561–564

Chapter 7
Power That Drives Circulation of the Atmosphere and Oceans

Abstract Flows in Earth's atmosphere and ocean are powered by the sun and the largest flow patterns are directly imposed. Earths atmospheric circulation is modified by instabilities that produce giant eddies that lead to large winds and weather at mid-latitudes. The result is three cells of vertical circulation in each hemisphere. Ocean flows near the surface are directly driven by heating and wind stress but the flows are modified by instabilities that produce large eddies. The vertical pattern of ocean circulation is modified by turbulent mixing.

7.1 The Atmosphere

Chapters 2–6 describe energy-flow structures driven by convection cells. The energy flow of convection cells in Chap. 2 proved to be like the energy flow in mantle convection modified by deep heat sources. A side benefit of mantle convection to all of us living on Earth is the organization of continents and ocean basins. The convection cells are also linked to the magnetic field by three-dimensional convection cells in the core. The central idea is that the form of the motion starts from instability and then becomes modified as the flows become greater. The form of the flows themselves is an energy-flow structure.

Earth's atmospheric and ocean circulation have large components that are directly forced by solar heating so the overall form is not completely an energy-flow structure. For the atmosphere, the vertical arrangement of the energy flow of the atmosphere is complex. Fist, the radiation occurs over many layers (Fig. 7.1). The layers occur because of absorption and radiation over a wide range of wavelengths. Second, the sunlight intensity is greatest at the equator and smallest at the poles so that each arrow in Fig. 7.1 varies with latitude. The flow of air in the atmosphere is forced directly by the variation of the buoyancy forces from the equator to the pole and vertical radiation balances. There is, however, a very substantial modification to the directly driven flow by instabilities that produce energy-flow structures. In the atmosphere, the largest modifications are perhaps the weather patterns and wandering jets and

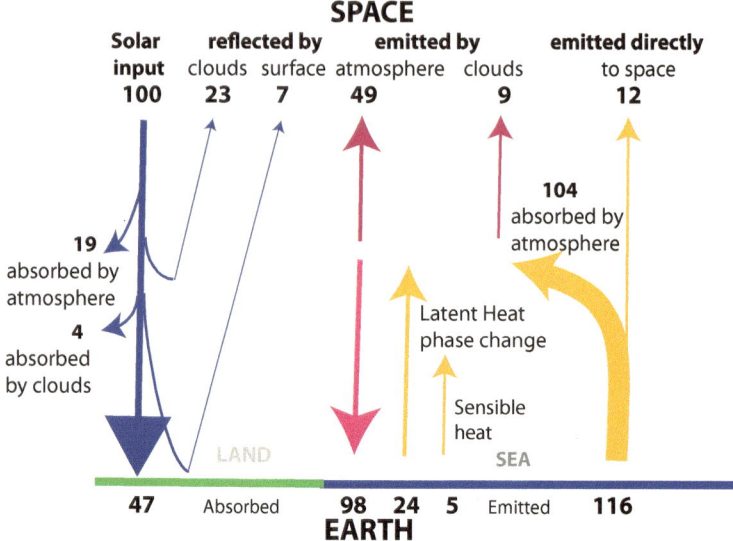

Fig. 7.1 The incoming and outgoing radiation flow for our atmosphere (NOAA public domain)

eddies at mid-latitudes. This chapter sketches how instabilities leading to these were discovered.

How large is the energy flow? To produce a perspective, the flow of heat coming up through the ocean floor and continents (geothermal heat) is about 5×10^{13} W and is thought to have remained at this value up to perhaps a factor of two or three greater during the entire history of Earth except for the early formation period over 4 billion years ago. This energy flow from inside the Earth is dwarfed by heat flow from sunlight entering the top of Earth's atmosphere at the rate of over 5.5×10^{16} W, which is three orders of magnitude (1000 times) greater (This is easily estimated using the value of the solar constant of 1.362×10^3 W/m^2 and multiplying by Earth's disk area using a radius of 6357 km). Naturally, not all solar radiation reaches Earth's surface. Some is directly reflected by cloud tops, ice, and the ocean surface, so it never reaches the ground. The total radiative energy that lands on the land and ocean surfaces to produce our climate, weather, and ocean surface temperatures is commonly accepted to be approximately 3.7×10^{16} W. This amount is still almost a thousand times greater than geothermal heating. Consequently, the temperature that we experience from weather and ocean flows are almost entirely driven by the sun and the radiation that returns the heat to space rather than from geothermal heating.

Let's check a consequence of this large heat flow rate with a simple calculation that I like to keep in my head. The oceans evaporate approximately one meter of water on average per year. This water must ascend into the atmosphere and return to the land and sea as rain or snow (At least, most of the return is by rain and snow precipitation; fog and dew give very small amounts). We find the total energy-flow rate using this water flux rate of an upward flow of water in vapor form of 3×10^{-8}

m/s, the latent heat of water for evaporation of 2.45×10^6 J/kg, a water density of 1000 kg/m^3, and an area of the oceans of 3.4×10^{14} m^2. This gives a total rate of heat of evaporation estimate of 2.66×10^{16} W. This is a large percentage of the sunlight heating the oceans, and consistent with the concept that a large part of the flow of energy in our climate involves evaporating and distributing water. We must remember that this estimate is an order of magnitude estimate and is rough. When divided by total heat flow, the ratio is somewhat larger than the ratio between received radiation and latent heat shown in Fig. 7.1, which is close to ½.

Evaporation rate, and hence the amount of water in the atmosphere is largely governed by the sensitivity of the vapor pressure of water on temperature. Indeed, Earth with a surface at the freezing temperature of water would experience virtually no evaporation, and incoming radiation would have to be balanced by back radiation that would not be hindered by water's greenhouse effect. Models of a cold ice-covered Earth are stable because of this. The reflection by ice returns the sunlight to space and some models produce a snowball Earth even with present conditions. There is evidence that snowball Earth existed for hundreds of millions of years before 700 million years ago. Fortunately, our present Earth is also stable and not easily pushed to a snowball Earth. Thus, Earth is bistable. It has two possible states, the snowball state, and the present state. Each state is locally stable. If the average present atmospheric and ocean surface temperature were slightly lower, then the evaporation rate would be smaller than at present, and evaporation removal of ocean heat would decrease. The ocean would thereby be less cooled by evaporation, get warmer, and restore that atmosphere to its present value. The warm deviation is likewise stable, and the present state is stable to small disturbances. This consideration includes the ocean but does not include all the complicated processes for the back radiation of the atmosphere in models that are presently used for climate studies of Earth, so this simple explanation of why Earth is at roughly our present temperature is one of many factors for climate studies.

How does the energy flow of sunlight affect our climate and weather? For our climate, the average temperature of Earth is a result of radiation received from the sun and radiation that returns to space from the warm land, the oceans, and the atmosphere. This balance is well studied but not easily understood in detail because many substances in the air have important effects on radiation. First is water in liquid droplet form, which makes clouds that vary greatly over time and location. Clouds directly affect the reflection of incoming sunlight. They also serve as a blanket to outgoing radiation. Second is water vapor in the air. It also serves as a blanket to outgoing radiation and is the greatest greenhouse gas. Water vapor is largely responsible for Earth temperatures being at present values. Third, water is one of the only substances present in vapor, liquid, and solid form on Earth. Describing and quantifying the energy-flow rate involving these three forms alone can easily fill an entire book. Fourth, other gases such as carbon dioxide, methane, and ozone also have greenhouse effects. Fifth, aerosols affect radiation balances. They are tiny particles other than water drops. They arise from living organisms, ocean waves (salt particles), volcanos (dust), and human industry (many different particles). Numerous books, reports, and scientific studies are written about the radiation balance of Earth, so we skip giving a detailed account.

In terms of our wind and weather, let's move beyond the energy and radiation balances and focus on the flows that the energy source produces. The atmospheric circulation and associated bands of high and low pressure on Earth's surface are shown in Fig. 7.2. The preferential heating near the equator, rather than at the poles, produces warmer temperature in the tropics and colder temperature near the poles. The global temperature difference would be expected to produce a general rising of the atmosphere in the tropics and sinking at the poles. Such a large-scale flow is directly driven and hence it is not an energy-flow structure. Atmospheric circulation on Earth does not have this simple rise-descent pattern. Instead, there are three overturning cells. Near the equator, a circulation pattern, called the Hadley Cell, has rising flow in the tropics and sinking away from there. This cell extends up to about 30° north and south and the mechanics is well studied. The next cell is the Ferrel cell, and the polar cell occupies polar regions.

Figure 7.2 indicates that the Hadley cell has flows both in a north–south direction and around the planet. Flow around the planet arises because of Earth rotation. The Earth surface is moving west to east at the fastest speed near the equator because that is furthest from the axis of rotation. Rising air from heating in the tropics requires converging to the Equator for flow near the ground and diverging flow away from the equator at higher elevation. The flow near the ground experiences surface friction as it moves toward the Equator. The air is also moving slightly further away from the axis of rotation and so it lags the rotation. This produces winds moving opposite

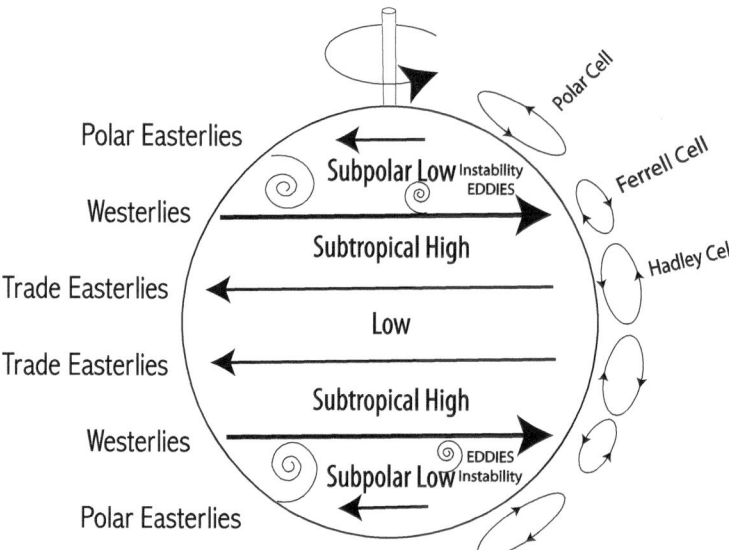

Fig. 7.2 The surface winds the pressure and the three pairs of overturning cells of Earth (NASA image)

in direction to Earth's surface velocity from rotation. This wind belt is called the easterlies, or the trade winds. Since air in the tropics rises, the upper air must spread north and south away from the equator and move toward the axis of rotation. The air has less friction with the ground, and from the conservation of angular momentum, the winds in the upper part of the Hadley cell further away from the equator than the trades move in the direction of rotation. This produces the westerlies, or winds blowing toward the east, also shown in Fig. 7.2. They are the location of the famous jet streams. Since the upper region has much less friction, shear between the upper and lower levels is greater and greater as one moves away from the equator. The consequence is that the westerly velocity becomes increasingly rapid away from the equator and an instability develops (Held and Hou 1980; Held 2000).

Although the explanation of the westerlies using conservation of angular momentum was known in the early twentieth century, the fact that the westerlies reach a peak and then decrease beyond 30° N and S was a puzzle. Before 1947 and 1949, when I was 6–8 years old, there was little progress toward understanding why the westerlies decrease beyond 30°. Although some models predicted shear instability of fronts associated with jets, it was not understood how the atmosphere breaks up into the large cells associated with weather patterns, storms, and cloud clusters. Then, Charney (1947) discovered a new type of flow instability for layered flows in a rotating fluid! He started with a set of calculus equations that were accurate for our atmosphere and showed that the equator to pole temperature distribution and an associated vertical shear in the winds that circulate around Earth can produce a growing disturbance consisting of waves propagating around the planet. The mathematical approach resembled Rayleigh's stability analysis by calculating the growth of a tiny perturbation. Although the mathematical solution is quite complicated, in general the unstable cells in that analysis are waves of pressure disturbances that propagate around the planet that are unchanging in a north–south direction.

In a 1948 Ph.D. thesis, Eady (1949) calculated stability with a simpler configuration. Because the model is so simple and the instability is so important, a brief description is useful here. Fluid occupies a channel extending to infinity in the east–west direction in a steadily rotating frame of reference. The axis of rotation is aligned with the direction of the gravitational force, and both are vertical. Therefore, it is a model of the atmosphere at mid-latitudes. There are no terms corresponding to the spherical shell. The outer and inner walls of the channel are hot and cold and produce a gradual temperature change across the channel. In a rotating fluid this makes the speed change in a vertical direction. The relation between a lateral temperature change and vertical shear for a rotating fluid is called *the thermal wind relation*. This fundamental balance exists for numerous flows in the atmosphere and ocean. It does not even invoke the distance to the rotation axis that is used to explain the easterly and westerly winds earlier in this chapter. In Eady's model, the thermal wind represents a region between the trade winds toward the west near the equatorial surface and the jet streams toward the east in the upper atmosphere. The mathematics of the instability is like the one found by Rayleigh in Chap. 2 in one sense because a very tiny perturbation grows. It differs in another sense because the growing perturbations are moving waves of temperature, pressure, and velocity variation. The instability itself is called baroclinic instability because the basic flow has a density change along constant pressure surfaces. The density field is stable to Rayleigh–Benard instability because denser fluid lies below lighter fluid. The growing unstable waves release some potential energy anyway. Energy for instability in our atmosphere is available from the tropics to pole temperature change. The growing cells have hot air rising and cold air sinking. This type of instability became a fundamental new contribution that has a continuing profound influence on theoretical meteorology and oceanography.

Between 1950 and 1960, laboratory studies with the colorful name of dishpan experiments gave tangible pictures of the unstable thermal wind. The baroclinic instability of the basic current and the eddies resulting from the growing perturbations are completely obvious. The experiments were initiated by Dave Fultz using two hemispheres with a gap in between containing water, with results published in 1950. The apparatus was simplified by Raymond Hide in a Ph.D. thesis published in 1953. A tank had two circular vertical walls to form an annular channel that was placed on a turntable rotating at a constant rate. Fluid, usually water, fills the annular channel. A common and simple demonstration for classrooms uses an outer cylindrical dishpan with the wall kept at room temperature containing an inner pan filled with ice water on a rotating turntable. Water occupies the annular region between the outer and inner walls. In the rotating frame, the flow near the warm sidewall is in the direction counter to rotation like the easterly trade winds in the tropics. The flow near the cold sidewall is in the direction of rotation like the mid-latitude westerlies. In the middle of the annular channel a vertical shear of the currents obeys the thermal wind relation. Flow is counter to rotation near the top and flow and is in the rotation direction near the bottom. For certain rotation rates, an instability is visible as a meandering of the center of a shear region. The eddies between the meanders are like the low pressures associated with storms and high pressures associated with clear weather on a weather map. Many of the annulus experiments like Fig. 7.3a produce wonderful images.

Although annulus experiments have been used to define regions of stability and eddy sizes for many years, newer approaches permit the measurement of energy flow. The eddies that are produced by baroclinic instability are fundamental to heat flows in our atmosphere and in the Antarctic Circumpolar current in the southern ocean. The experiment in Fig. 7.3b accomplishes the transport mechanism for the southern ocean. The basin is cylindrical filled with water with no inner wall. Pumps and a rotating lid produce flows away and toward the axis, so the buoyancy flow can be measured directly. Dyed water is pumped in from the outer wall and ordinarily would continue to spiral and fill in, but instead eddies transport some of it away from the circular current. After many hours, the eddies have come to a steady equilibrium. Not only does the instability produce the eddies, but the eddy indicated in the figure shows the transport mechanism. One eddy separates off from the frontal region to complete transport from the dyed dense water to the clear lighter water. The saltwater budget is equivalent to a heat flow budget, and this shows directly that the heat flow is accomplished by detached, wandering, and isolated eddies. Detached eddies like this are widely found in oceans. Perhaps storms in the atmosphere serve the same purpose.

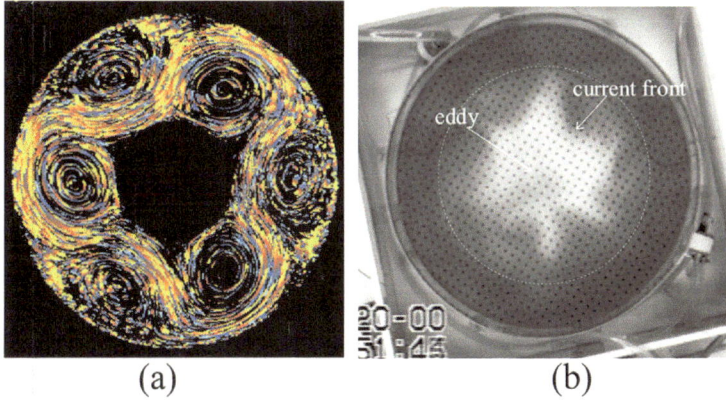

(a)   (b)

Fig. 7.3 **a** Streaks of particles showing eddies in an annulus experiment. Photo produced by Stephan Risch and supplied by Peter Read. **b** A laboratory experiment that quantitatively measures the flow of dense water (dyed) below clear water. This is driven by inflows of fresh clear water and dyed salty water. The pumping includes the removal of fluid through the holes that make up the perforated base of the tank. A rotating lid spins the clear fluid. The grayscale is an indication of the height of the dense layer. After many hours, the eddies have come to a steady equilibrium. The eddy wandering away from the meandering front removes dyed salt water from the current around the tank and completes the saltwater budget (Cenedese et al. 2004)

When extending this model to Earth with either a more complicated theory or using a numerical model, one often includes the spherical shell geometry that is possessed by our rotating atmosphere as many early meteorologists like Charney and Fultz originally intended. For an atmosphere that is thin compared to the planet radius, this makes the effect of rotation to be a maximum at the poles and zero at the equator. This equator to pole difference is frequently modeled as a uniform change in rotation rate from north to south, and for theoretical discussion this is called the beta effect. Present laboratory experiments cannot duplicate the beta effect for flows with vertical density changes, but numerical models and other experiments can, which is fortunate because the beta effect is very important for numerous reasons.

In the atmosphere, Rossby waves (first analyzed in the 1930s) exist from the beta effect. They help to limit and propel the westerly jets like the jet stream near the northern terminus of the Hadley cell of Earth. The waves also help to make a band of atmospheric sinking at mid-latitudes rather than near the poles (Often there are two jet streams, one nearer to the tropics and one nearer to the polar circulation). The angular momentum transport from north and south of the jet stream regions is also accompanied by equator to pole heat transport that is liberated by the baroclinic instability with the beta effect (Starr 1968). Angular momentum is transported north and south not only by Rossby waves (Thompson 1971) but by energy flow in general (Whitehead 1975). We conducted laboratory experiments other than the dishpan variety to show these effects (Figs. 7.4 and 7.5). Therefore, the meandering and eddies are energy-flow structures that help to govern our climate.

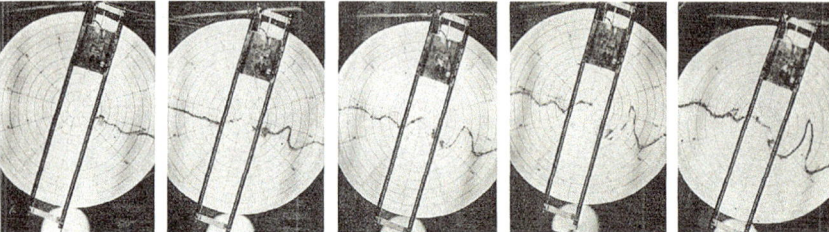

Fig. 7.4 Dye streaks in water on a turntable showing a current generated by the energy source. A paddle is under the square plate and the photo in each frame shows a jet halfway between the center and rim to the right of the frame. Photos are taken every 16 s. The rotation is counterclockwise and the current toward the energy source is balanced by countercurrents at other radii

Fig. 7.5 This beta boat stirs water in a laboratory beta plane. It not only propels itself but also produces flow and counterflow around the rotating tank (Waterman 2009)

The polar cell has sinking near the poles that must be balanced by rising some-where away from the poles. Since the energetic eddies in the jet stream region have a sinking component, a second cell called the Ferrel cell must exist in between the Hadley and polar cells. The Ferrel cell is variable and generally weak compared to the persistent trade winds and polar easterly flows. It has a second polar front like the jet stream that is supported by eddies that are less energetic than the weather systems further toward the equator. Occasionally, all cells possess very strong storms.

The atmosphere contains many smaller energy-flow structures such as cumulus clouds, thunderstorms, tornados, hurricanes, dust devils, dust storms, lightning, and certain frontal structures, to name a few. All of them are easy to recognize because they have the aspects of energy-flow structures we already have seen. The energy production that leads to their structural forms is balanced by dissipation, and they vanish when the energy-flow feeding them vanishes.

7.2   The Oceans and Their Surface Circulation

Ocean surface currents are well-known; maps like Fig. 7.6 have been developed and used by sailing captains since the ocean surface was first explored. The map is deceptive. Although many of the arrows in Fig. 7.6 are correct in principle, they generally refer to a drift. Quantitative mapping of the average surface flow and the time-dependent fluctuations has only become possible after satellites had extensive global coverage. Before satellites, the drift could be calculated using data sets of temperature and salinity along vertical sections taken by oceanographers and using the thermal wind relations to calculate currents. Although widely used, the thermal wind calculation requires a value that must be added to any profile of velocities at every elevation (mathematically we call this an integration constant). Historically, the calculations used an assumption that was called the "depth of no motion" that was assigned to have zero velocity at some great depth like three or four km. Early on, that assumption was considered acceptable, but with the development of moored current meters, cases were found where that assumption is incorrect. Presently the calculation is done in conjunction with currents that are either directly measured at some depth or measured at the surface by satellite altimeters.

There is production of energy by the wind work, and a great deal of attention has been directed toward the wind-driven currents in the ocean. The currents are most rapid in the upper kilometer. The ocean water in regions up to about 50° N and S have a surface layer of water with a temperature warmer than 10 °C that is roughly 2 km thick called the thermocline. The circulation in this layer is particularly energetic, and it includes many of the currents in Fig. 7.6. These include east–west currents in equatorial regions, large gyres driven by the wind called subtropical gyres north and south of the Equator in most oceans, the Antarctic circumpolar current, and subpolar gyres (in the North Pacific and North Atlantic). All are driven by wind stress. The energy received by the ocean surface is dissipated in various ways such as: in a turbulent surface mixed layer; in surface wave production; in surface wave dispersion; in surface wave dissipation; in the drag on large surface currents; by internal waves; and by mixing down into layers of cold denser water. The wide range of dissipation mechanisms makes detailed energy-flow budgets difficult to measure globally. A notable exception is in the paper, "The Work Done by the Wind" (Wunsch 1998). (This paper calculates work rate, even though the word "rate" is occasionally

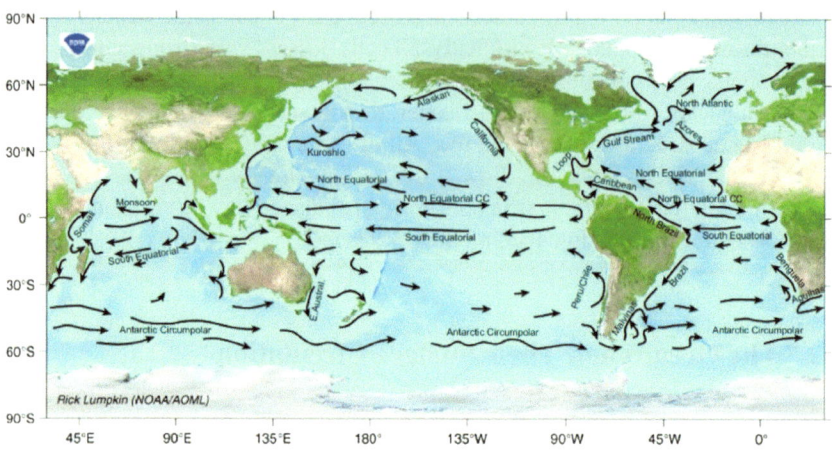

Fig. 7.6 Surface currents of the global oceans. Image by NOAA public domain

dropped.) Using satellite estimates of wind stress and surface speeds of the ocean globally, the rate is calculated to be 3.3×10^{11} W.

Current meters developed for operation periods of 6 months to 2 years duration were developed in my department just when I arrived at Woods Hole. Internal waves and tides with periods of hours and days became well documented, but the data also contained large slow (months or longer) time-dependent fluctuations with poorly known origins. There are two principal sources of these fluctuations. The first are simply time-dependent changes. They can occur yearly and strengthen and weaken from a variability of wind stress. The second is eddies. The first maps of such eddies were probably obtained as a time-history of a Gulf Stream ring in the 1950s. At first, these were the only eddies documented. A coordinated large program called MODE in the mid 1970s made the first map of a deep eddy in the open ocean. Satellites produced maps of ocean temperature soon after that, and within 10 years it became known and completely accepted that the ocean surface is full of such eddies. A dramatic photograph of many eddies at the surface is shown in Fig. 7.7.

7.3 Deep Ocean Energy Flow

The oceans lie under the atmosphere and are subjected to various flows of energy.

We listed in 7.1 the equivalent energy flow leaving the oceans in the form of a change of phase of water from liquid to vapor by evaporation with the energy rate of 2.6×10^{16} W and noted that this is more than 2/3 of the sunlight heating rate that drives the weather and oceans. Since the oceans cover roughly 2/3 of Earth's surface, it seems clear that a balance between radiation and evaporation is important for Earth.

Fig. 7.7 Eddies on the ocean surface revealed by various colors that represent different temperatures (data from satellites, a NASA image, public domain)

The cycle of energy flow of up through the atmosphere that starts with the production of water vapor by evaporation and ends with its return to the surface by precipitation is interesting and the magnitude is large. In contrast to this, the large currents inside the global ocean are governed by the thermal wind relation, and the contribution to salinity variation is small. Larger salinity contributions are found in some regions such as the Arctic Ocean and in continental shelves. The temperature range of our present oceans lies from the tropics with a temperature lying in the range from 20 to 30 °C to the polar ocean water temperature that is close to 0° and nearly freezing. Most of the deep ocean water has descended into the deep in polar regions and spread out to fill the oceans. The global average ocean water temperature is close to 4 °C. The effect of high pressure on thermometers before the mid-1800s caused incorrect readings that led to ideas of high deep temperature. This confused the issue, but after correct pressure protection was adopted for deep ocean thermometers by 1900, we know that the water near the polar regions sinks to the bottom. It spreads out through the basins simply because it is colder and denser than thermocline water. An estimate of the potential energy release rate of sinking in the polar regions from the temperature difference is made later in this chapter. Its magnitude is useful to appreciate because it is one order of magnitude smaller than the freshwater-salinity cycle and four orders of magnitude smaller than the power in the evaporation rate calculated above.

Although the water cycle within the ocean is small, localized variation of ocean salinity introduces a significant complexity to the location of a sinking region. Tropical oceans generally have the highest evaporation rates and precipitation. However, precipitation locations are patchy and ocean salinity is likewise patchy. Salinity is also patchy because some basins around the ocean have evaporative basins that export saline water locally. Two examples are the Mediterranean Sea and the Red Sea. In addition, the locations of freshwater supplied to the ocean are dominated by rivers. The surface salinity of some of the polar ocean water has the lowest value of all the large ocean basins. The Arctic Ocean is a good example of a basin with no deep sinking because of a layer of lower salinity water hundreds of meters deep. In fact, deep sinking is not generally common in polar regions, and it occurs only in basins near the sides of polar regions is water with relatively high ocean salinity is present.

Worldwide, the deep ocean sinking locations tend to be places where salty surface water gets a deep mixed layer in winter. To make it even more complicated, the times and locations with such deepening are episodic depending on the severity of winter storms. Personally speaking, in the 1970s the old timers at WHOI told me stories about deep convection events in the Labrador and Greenland seas but such events were not occurring there anymore. Then, in the 1980s, deep convection activity picked up. Deep mixed layers occurred in the Norwegian and Greenland Seas, and the water spread out later. Therefore, the Labrador Basin of the North Atlantic, the Norwegian Sea Basin, and the Greenland Sea Basin all have many years with and without sinking. This is also true in certain locations near Antarctica. There are further complications, because some of the very coldest water in Antarctica accumulates at a few hundred meters depth under a glacier where the liquid sea water is so cold that it would freeze if conveyed to the surface. Because of the localized and, in some cases, episodic nature, it is presently challenging for oceanographers to quantify the volumes of new cold water that sinks to great depths each winter. Also, it is difficult to map out the amount of water that is not reheated in the following summer and difficult to describe the flow paths of the cold dense water over the next few years when it flows into the main ocean.

Another complication is that the coldest and densest water accumulates in smaller basins and then flows through gaps into the larger ocean basins. Flows through the Denmark Straits and the Faroe Bank Channel bring some of the coldest water into the deep North Atlantic. Flows through the Filchner Depression near the Weddell Sea bring the coldest water into the Antarctic Region.

Despite all the complications, it is clear that cold water fills many local basins and then spreads out. Some of the spreading of the deep water in the Atlantic can be visualized using Fig. 7.8. It is a map of the locations of water colder than 1.8 °C potential temperature in the deep Atlantic (Potential temperature is a term used by oceanographers for a correction of temperature due to ocean depth-pressure changes). The figure includes the contours of the depth below the surface with the potential temperature at that value. An atlas showing many different constant potential temperature surfaces like this are in a technical report by Worthington and Wright 1970. The figure influenced the understanding of deep ocean circulation by many people (including me) in the mid-1970s. The figure indicates that cold water occupies several gaps and basins in the Atlantic and occupies two tongues. One extends from the north, and the second extends from the south. The coldest water extends up from the south in Fig. 7.8 and is due to water that flows northward from the Antarctic region called Antarctic Bottom Water. It flows through a gap near the equation into the North Atlantic where it ends. Water colder than 1.8 °C potential temperature lies below this surface, and the isothermal tongues terminate further south. It is almost inescapable from this figure to conclude that the water from the Antarctic region flows northward and is mixed slightly as it moves. This view has been verified and quantified by current meter measurements in two locations where the gaps concentrate the northward flow. One gap is at approximately 30° S and the other is near the equator. My wonderful colleagues Mindy Hall, Mike McCartney, Dick Limeburner, and Claudia Cenedese, and I have measured the flow rate at the passage near the equator. In two periods of time, the rate is almost the same with a rate of 2.0×10^6 m^3/s both in the periods 1992–1994 (Hall et al. 1997) and in 1999–2004 (Limeburner et al. 2005).

The other tongue, extending down from the north, signifies water that oceanographers call Lower North Atlantic Deep Water. This flow starts from water that is very cold and dense in the Greenland-Norwegian Sea. Cold dense water flows southward through the Denmark Straits and descends into the North Atlantic Ocean. Two jets north of Iceland feed into the current. They are the East Greenland current, and the North Islandic Jet. The overflow water spills down to 3000–4000 m depth along the flanks of Greenland's continental rise and mixes as it sinks. Then, the tongue extends southward along the flank of the North American continental rise before terminating. Overflow volume fluxes are summarized in Pratt and Whitehead (2007).

Waters above these tongues overlie each other and mix. A larger water mass called North Atlantic Deep Water is comprised of upper and lower layers from many deep-water sources. It accumulates in the Labrador and Greenland Seas of the North Atlantic, flows southward in western boundary currents within the Atlantic Ocean, and ends up circling the world. Parts are swept up into the Antarctic Circumpolar Current, and parts are concentrated in the western boundaries of the Pacific, and

Fig. 7.8 The depth of water is colder than a potential temperature of 1.8 °C in the Atlantic (contours of depth are every 500 m over the shaded bins). The thin contours in unshaded areas are the 3000 m bottom contours (adapted from Fig. 1.6a in Pratt and Whitehead 2007)

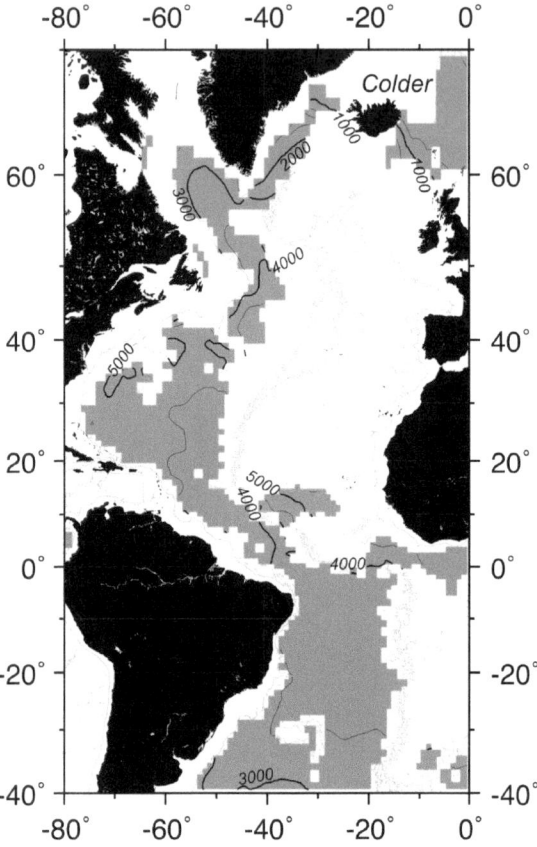

Indian Oceans. These deep and bottom waters comprise sources for a large percentage of the water in the deep ocean.

Figure 7.9 is a sketch of the vertical circulation of the Atlantic Ocean. We need to point out that such cross sections can be deceptive because the actual circulation is three-dimensional. For example, in some basins, ridges intersect the middle and each side is different. The Pacific and Indian Oceans are similar in most aspects to the Atlantic, except that there is no deep sinking in the North. The deep current that feeds almost all the deep water into the deep basin of the North Pacific Ocean flows northward near Samoa and then seems to join concentrated currents along the western boundary current within the North Pacific Ocean basin. The Indian Ocean also has no deep sinking in the north. so the inflow of deeper waters into the basin is only from the south. There is some evidence of concentrated currents along the western boundary of that ocean, but the western sides of that basin contain both Madagascar and the Southwest Indian Ridge, and the detailed pathways along the complex continental rise are not known. The amount of mixing upward in the Indian Ocean might be less than in the Atlantic and Pacific, but it is not quantified.

Fig. 7.9 The vertical circulation pattern for the Atlantic Ocean

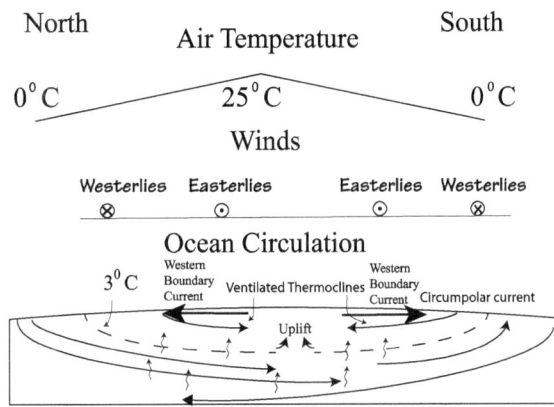

Water with a temperature above 3 °C lies above the dashed line in Fig. 7.9. It is directly in contact with the atmosphere and has a more complex wind-driven flow. That flow is almost perfectly horizontal with only a small vertical component (Pedlosky 1987). It is characterized by slow mean circulation patterns in the form of gyres, along with concentrated currents, particularly along western boundaries, and eddies. A localized aspect of circulation within this water is called the ventilated thermocline. There, water flow is characterized by circulation along contours of constant density that is fed at the upstream end by the winter mixed layer. Ventilated water sinks along curved trajectories and is then fed into the wind-driven gyres that feed the surface western boundary currents, such as the Gulf Stream and Kuroshio.

The complete pathway for any water that sinks must include a return path to the surface, and this requires some mixing of the cold water with warmer water. The locations of upflow and return are not known, but some recent evidence shows some systematic rising of deep water around mid-ocean ridges. A detailed study to see the dynamics of the mixing in a canyon in the Northeastern Atlantic was recently reported by Wynne-Cattanach et al. (2023). Their documentation of flow up the canyon is the first direct observation of the rising that might potentially balance the sinking of cold water in polar regions. The actual patterns of deep ocean flow are not known in detail, although a large component of flow is concentrated in currents along western boundaries that were theoretically predicted in the 1950s. The currents are measured in many basins but reconciling them with tongues of water as in Fig. 7.8 has not been successfully done. The overall flow between basins has been described as a conveyor belt, but there are many layers for the conveyor belt flow. Each layer extends into a basin where the water is mixed up with warmer water above it and vanishes as a distinct layer. This entire circulation pattern is known as thermohaline circulation, and because the oceans have multiple basins, the pathways of sinking and returning to the surface are complex and often simply sketched out or not even known. Popular pictures are very misleading in some cases, as the actual migration paths of water are not easily drawn because the water mixes as it migrates.

The energy cycle for ocean convection involves the vertical circulation patterns and these differ from the horizontal circulation patterns of wind- driven flow studied by most physical oceanographers. Figures 7.8 and 7.9 show that the sinking of dense water occurs in polar regions north and south of the Atlantic Ocean. The cold dense water accumulates near the bottom and spreads out. There are a few different layers since some water sinks in the Northern Hemisphere and other water sinks in the Southern Hemisphere. The upward return flow is partly driven by the Antarctic Circumpolar Current that lifts water because of Earth rotation. This is indicated on the right of Fig. 7.9. Other vertical flow up comes from turbulent mixing. This important turbulent process is measured in only a few ocean locations so far. The exact intensity, locations, and times of all the mixing sites are still unknown in detail although there is some indication of intensification occurs near rough bottoms such as mid-ocean ridges. The deep circulation that is sketched by the jagged arrows in Fig. 7.9 is related to the turbulent mixing of the water.

We conducted a laboratory demonstration to illustrate the relation between convective sinking near the dense water source regions and turbulent mixing (White-head and Wang 2008). The apparatus is shown in Fig. 7.10. A tank of water has two sources at the two opposite top ends, with one supplying fresh water and the other supplying denser salty water. Both are pumped in with the same flow rate. A spillway in the middle of the top allows the water to leave, so that the total volume of the water in the tank stays constant. Mixing is provided by a towed vertical rod that moves quickly enough back and forth to shed small-scale turbulent vortices it its wake. The mixing intensity can be varied for each experimental run, but it is always large enough to make turbulence (the Reynolds number is bigger than 1000), but

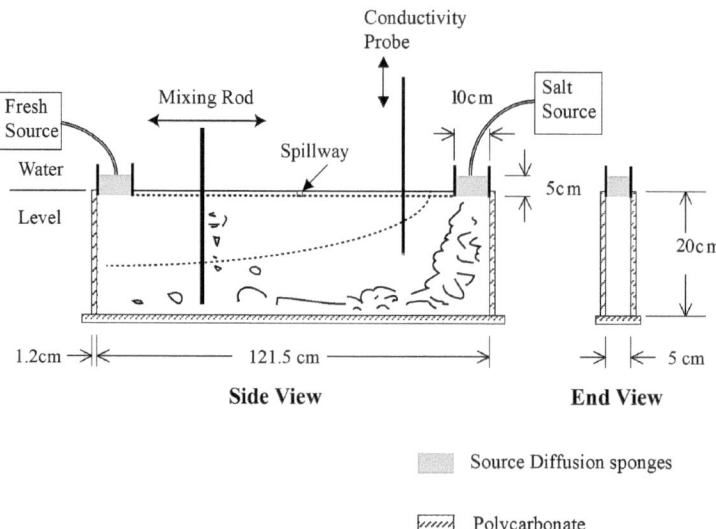

Fig. 7.10 The deep circulation cell driven by turbulent mixing

small enough to not mix the water all up at once. In terms of energy, the flow induced by the rod, although turbulent, does not have enough kinetic energy for a parcel of the bottom salty water to penetrate all the way up to the surface and for a parcel of fresh water to penetrate all the way down to the bottom. Without rod motion, there is no turbulence. Mixing is minimal and the freshwater occupies a shallow layer flowing to the spillway. Below it is undiluted salty water filling the rest of the basin. This water also flows to the spillway (This situation is like the Arctic Ocean). The purpose of the experiment is to document the effect of the mixing rate on the salinity distribution of the water. The first effect is that with the rod steadily moving back and forth, the fresh water coming in at the left is mixed downward by turbulent eddies. In this case, the water at mid-depth becomes a mixture of fresh and salty water, but the salty water from the source on the right is denser than this mixed water so that it continues to sink to the bottom. To complicate matters, the sinking plume becomes turbulent, and this also draws down some of the mixed water. The technical word for this mixing of water into the sinking plume is *entrainment*. A large overturning circulation cell is introduced by inflow into the turbulent plume and the rate of total overturning can be many times greater than the rate of the two source pumps. In summary the total mixing in the tank is produced by both the rod mixing and by the mixing of the turbulent plume.

The turbulence intensity of the sinking plumes has a direct and complex effect on the overturning circulation and on the final density profile, but there is one simple fact for this experiment. After the salinity distribution of the bulk of the water has become steady, the fresh water that is moved downward by turbulence from both the mixer and the turbulent plume (white patch in Fig. 7.11) must have the same volume flow rate as the volume flow rate of the original sinking salty water (black patch in Fig. 7.11). The mixed water has the density of the water leaving the tank at the surface, so imagine that the white and black patches mix and then rise to the surface and leave the tank, and the density profile of the interior is unchanged. Therefore, the fact that fresh and salty water volume flow rates down into the interior are equal means that the energy-flow rates of dense water sinking and the mechanical work rate to move down surface water by mixing are the same. This is true at every depth where salty and fresh mix together, but this is not the complete energy balance of the experiment (or the ocean), so caution is necessary. Mechanical work is done by the mixer to cause the ambient fluid to rise around the white and black volumes on

Fig. 7.11 Mixing down of freshwater by turbulence equals saltwater sinking down in the plume

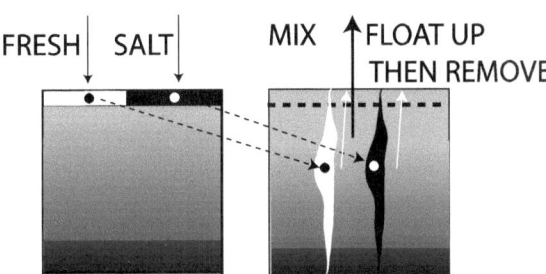

the right-hand side of the figure, but also work is done by the turbulent plume. The division of work done by each depends on the details of the mixer efficiency and the dynamics of the turbulent plume.

Let's apply the balance to the ocean. If the ocean is neither heating up nor cooling down, then the net vertical flow of heat in the deep ocean layers would likewise be zero, that is, heat conveyed down by turbulent mixing everywhere would warm the cold water supplied by sinking in polar regions (At least if we assume that the flow of heat up from the ocean floor is too small to matter for this approximation). The balance assumes a steady-state ocean, so any slow change of the mean temperature of the entire ocean basins is not allowed. Such a drift of the total mean ocean temperature is not yet measured, although the warming of small volumes of surface layers from global warming is now reported. Assuming that the balance is steady, or at least that it has been steady for many years before the recent global warming trend began, let's quantify the potential energy cycle of our deep circulation of Earth's oceans. Deep water formation corresponds to the dark salty water that sinks in Fig. 7.11. Mixing down of warm ocean water from the surface corresponds to mixing down the fresh water in Fig. 7.11. Recall that temperature, rather than salinity, dominates density for deep global ocean circulation and salinity differences only contribute a few percent to the total energy budget. The typical temperature difference between tropics and poles is approximately 20 °C. The coefficient of expansion of water averages $\alpha = 1.6 \times 10^{-5}$ so using the approximation for water density of $\rho_0 = 1000$ kg/m^3, the density difference between the hot surface water and the source water for deep flow is $\Delta\rho = \rho_0 \times \alpha \times \Delta T = 0.32$. The total volume flow rate of cold water descending to a great ocean depth such as 4000 m is not precisely known, but we can estimate that it would range from $Q = 6 \times 10^6$ m^3/s, which it the sum of known values of the Denmark Strait and Faroe Bank overflows (Table 2.14.1 in Pratt and Whitehead 2007) plus 10^6 m^3/s for the Weddell Sea, to a larger value of $Q = 2 \times 10^7$ m^3/s to account for sinking at shallower depth such as the Greenland Sea, the Labrador Sea, and in the Southern Ocean Given that the present deep ocean flows are steady, Fig. 7.11 implies that the both the volume flow rate and the potential energy-flow rate of the light warm water that sinks down within the ocean from mixing (that is known to be driven by energy from the tides, winds and sinking dynamics) is of equal magnitude to the volume flow rate and the potential energy-flow rate of unmixed cold water plunging down at the polar regions. Using the approximation that the acceleration of gravity is 10 m/s^2, then the total rate of potential energy release Er is given by the formula Er $= g \times \Delta\rho \times d \times Q$ (introduced below Fig. 2.12). Using $d = 4000$ and above values, the rate of potential energy released by dense water sinking in the oceans ranges from 7.7×10^{10} to 2.6×10^{11} W.

Can this be compared to other measurements? A significant percentage of the total energy dissipation in the deep oceans is from deep ocean tidal mixing. That magnitude is estimated to be approximately 2×10^{12} W (Wunsch and Ferrari 2004) from ocean data. This is between 25 and 8 times larger than our calculation of the rate of potential energy release. Why is it greater? One factor is that the sinking estimates are too low because the turbulence of the currents where the dense overflow water descends to thousands of meters is neglected. That turbulence is not tidal in origin.

Second, viscous dissipation is intense at mixed layers that occur at every ocean boundary. Third is that even in the stratified ocean interior, a sizeable fraction of the 2×10^{12} W of turbulent energy tidal dissipation is viscous dissipation and not the energy of mixing warm water down. Fourth, laboratory measurements give the ratio of the energy loss from buoyancy dissipation to turbulent viscous dissipation to be 0.2 but the range of uncertainly is high. Some cases exist with other values. Multiplying 2×10^{12} W by 0.2 gives 4×10^{11} W and this differs from values in the previous paragraph by less than a factor of five for 7.8×10^{10} W and a factor less than two for 2.6×10^{11} W. Agreement within such factors of such a crude calculation can be made a bit better with high-resolution data, but the project is too challenging for this chapter for many reasons. First, the ocean density profiles are uneven, with most of the warm water lying in the top km of the ocean rather than deeper. Second, there is other descent of cold water to the upper kilometer of the ocean from surface circulation driven by wind stress, particularly in the Antarctic Circumpolar Current. Third, the detailed measurements of turbulent dissipation in the ocean so far have values that vary from place to place, and complete coverage of all regions is extremely difficult and costly to perform. Fourth, seawater has a variable coefficient of expansion from salinity and pressure variation that is substantial and not included in this simple estimate. The fact that our estimates of dissipation lie approximately in the above ranges is a good indication that we are close to the correct values.

Returning to the theme of this book, which ones of the great ocean flows are energy-flow structures? It seems that within the atmosphere and oceans, some of the turbulent structures are energy-flow structures and others are not. The whirling storms in the atmosphere and many eddies in the ocean are produced and are maintained by energy flow. Many eddies resemble the eddies from the original dishpan experiments in Fig. 7.3. The growth phase of the eddies involves the flow, but their steady flow has almost no dissipation at all and has a balance between the Coriolis force of the rotating moving fluid and pressure. The source of the energy flow to make such eddies is complex, too. Some of it comes from the kinetic energy of large currents, and some comes from the liberation of gravitational kinetic energy of lateral density changes. To confuse the issue, such eddies seem to be a key component of two-dimensional turbulence as well, so their energy might cascade through different wavelengths with time. Finally, it is observed in both nature and the laboratory that large-scale eddies finally lose energy from processes that are very small, perhaps centimeters or less. Small-scale three-dimensional turbulence makes fine structures that dissipate energy by producing heat. The evolution of energy through different length scales is understood roughly but not in detail. Exactly where this occurs is not documented so far.

References

Cenedese C, Marshall JC, Whitehead JA (2004) Thermocline depth and exchange fluxes across circumpolar fronts. J Phys Oceanogr 34(3):656–667

Charney J (1947) Dynamics of long waves in a baroclinic westerly current. J Meteor 4:135–162

Eady ET (1949) Long waves and cyclone waves. Tellus 1:33–52

Hall MM, McCartney MS, Whitehead JA (1997) Antarctic bottom water flux in the equatorial western Atlantic. J Phys Oceanogr 27(9):19031926

Held IM (2000) The general circulation of the atmosphere. Woods Hole Oceanographic Institution Geophysical Fluid Dynamics Program, Woods Hole Oceanogr. Inst., Woods Hole, Mass. http://gfd.whoi.edu/proceedings/2000/PDFvol2000.html

Held IM, Hou AY (1980) Nonlinear axisymmetric circulations in a nearly inviscid atmosphere. J Atmos Sci 37:515–533

Limeburner R, Whitehead JA, Cenedese C (2005) Variability of Antarctic bottom water flow into the North Atlantic. Deep Sea Res II 11:333–346

Pedlosky J (1987) Geophysical fluid dynamics. Springer, p 710

Pratt LL, Whitehead JA (2007) Rotating hydraulics: nonlinear topographic effects in the ocean and atmosphere, vol 36. Springer Science & Business Media, 589 pp

Starr VP (1968) The physics of negative viscosity phenomena. McGraw Hill, New York, 256 pp

Thompson RO (1971) Why there is an intense eastward current in the North Atlantic but not in the South Atlantic. J Phys Oceanogr 1(3):235–237

Waterman SN (2009) Eddy-mean flow interactions in western boundary current jet. PhD dissertation, Woods Hole Oceanographic Institution-Massachusetts Institute of Technology Joint Program in Oceanography/Applied Ocean Science and Engineering

Whitehead JA Jr (1975) Mean flow generated by circulation on a β-plane: an analogy with the moving flame experiment. Tellus 27(4):358–364

Whitehead JA, Wang W (2008) A laboratory model of vertical ocean circulation driven by mixing. J Phys Oceanogr 38:1091–1106

Wunsch C (1998) The work done by the wind on the oceanic general circulation. J Phys Oceanogr 28:2332–2340

Wunsch C, Ferrari R (2004) Vertical mixing, energy, and the general circulation of the ocean. Annu Rev Fluid Mech 36:281–314

Wynne-Cattanach B, Alford M, Couto N, Drake H, Ferrari R, Le Boyer A, Mercier H, Messias MJ, Garabato AN, Polzin K, Ruan X (2023) Observational evidence of diapycnal upwelling within a sloping submarine canyon

Chapter 8
Speculations About Energy-Flow Structures in Nature and in Human Activities

Abstract The division of energy flow for basic human needs is described for three social systems: two primitive societies and the United States in 1955. Energy used for food ranges from less than half the available energy in all systems. Energy used to get water is less than 5%. Little energy is used for religion. Estimates of the energy used for shelter vary.

8.1 The Use of Energy Flow by Social Groups

The flow of energy is obviously complicated in most natural situations. In living cells, energy flow maintains the flow of nutrients and fluids, but each cell has its own unique chemical composition, energy flow and history. Every living cell in nature is evidently born, lives, and then dies. It consumes energy during life and dissipates some of it. The manner of energy consumption and dissipation continues after death but differently and leading to decay. Although certain seeds and spores live very long times, starvation generally leads to death. The difference between a living cell and the convection cells in Chap. 2 is that the living cell is governed by complicated genetic instructions that are passed on to other cells during cell division, whereas the convection cells have no such instruction codes.

We can thinks about how to make a convection cell that obeys instructions from a memory. Let us consider the case of cellular convection in a layer extending in two lateral directions with a few dense small solid particles lying along the bottom. If the vertical energy-flow stops, the Rayleigh number becomes zero and the cells go away, but when energy flow starts again, and the Rayleigh number exceeds a critical value Rac, then perturbations grow. In this case, for certain values of the drag and density of the particles, the arrangement of the solid particles will dominate the growing perturbation compared to random noise. We saw in Fig. 2.11 that above Rac there is a range of wavelengths for roll cells that can exist. The bottom particles can provide instructions about their wavelength and the orientation of the axis of the rolls in the two lateral directions.

Social systems involve a great variety of complex and complicated interactions, and in general, they don't obey rules that govern sciences like physics, chemistry, and biology. However, social systems possess many quantifiable commodities such as money, gold, ribbons, and fuels. The trouble is that most of these commodities from a scientific viewpoint is that they can be replaced with other commodities. That is, they can be interchanged. Energy flow is obviously also a quantifiable commodity, and it's not easily replaced by anything else. Therefore, its conservation might be obeyed by a social system in the same sense as the examples in this book. The practical problem in thinking of energy flow as a conserved property is that the energy flow is complicated and sometimes even profound. For example, to produce a drinking cup in our modern culture involves many steps: the use of heat flow to melt raw materials; heat flow for the preparation of some chemicals; the consumption of gasoline or oil to construct warehouses and stores; fuel consumption for transportation; growing some food for the working personnel; and even shopping. Possibly a good approach is to look at large systems. The energy use by several social systems has been studied. Generally, in such studies the word energy is used rather than energy flow, but studies generally consider a specific amount of time that energy is used, so energy flow rather than energy is studied.

I did a project back in the 1980s (Whitehead 1987). I was able to look at 3 social systems and see how their energy was used for their basic human needs. It never occurred to me that energy flow was important rather than energy itself, but all three were for a specific time interval, so energy flow was the basic variable. The social systems were two aboriginal groups in Australia when they were living naturally, as well as the United States in 1955. The data indicated several things. First, the percentage of energy spent on food varied by about only a factor of 2 or at most 3 between the simple subsistence groups and the complex postwar economy of the United States. It was a surprise, because I thought that a subsistence culture would spend much more energy gathering food. Therefore, I used the results to hypothesize that the range of the available energy used for food is from 15 to 45% universally. Second, less than 5% of the energy was used to acquire water. In view of the fundamental need for water, this was also a surprise. Third, more energy was used for heat than for abode construction. Error bars are large as one might expect, but heat and constructing a single abode have a significant difference between energy flow and the total energy for construction. A supply of heat is really a supply of heat flow that is needed continually, but an abode can take a lot of energy flow to build over a short time and then be used with little energy flow later. A large population studied over a long time would smear out the difference between heat flow needed for warmth and heat flow needed for shelter. Fourth, a very little portion of energy was used for religious or cultural activities. Finally, the transfer between thermal energy and mechanical energy was different. The aboriginal cultures had very different uses for thermal energy than they did for mechanical energy. In contrast, the United States in 1955 had a clear overlap between thermal and mechanical energy since machines could convert thermal energy to mechanical work and vice versa. I speculated that energy technologies that result in more efficient use don't change the percentages used for different human needs but change something else, such as food production and a larger population.

Reference

Whitehead JA (1987) The partition of energy by social systems: a possible sociological tool. Am Anthropol 89:686–700

Chapter 9
A Tutorial Review, Simple Energy-Flow Calculations, and Basic Definitions

Abstract This gives a brief history of the tutorial of simple mechanical energy and its conservation along with the definitions of kinetic and potential energy, and then a simple example of the rate of production of energy for buoyantly driven fluid flow. The quantification of energy flow is reviewed along with explanations of orders of magnitude and dimensionless numbers. An example is given of a fluid flow driven by energy flow, notably a water flow driven by the heat flow into a tube below a lake leading to a hot spring.

9.1 Energy and Work

Here, we review my understanding of simplified energy and work. Some people will know most of this and needn't bother. Some others might know this material, but this might help them to teach the public or give a course to students not majoring in science. Others will find this instructive.

We start by reviewing some fundamental knowledge about energy and work. Later, we will extend the knowledge to the flow of energy and the rate that work is done. Energy occupies our universe and is conserved just like mass is conserved (I ignore the small rates of conversion between energy and mass expressed by the famous relativity formula $E = mc^2$). Our understanding of work and energy (and other concepts such as momentum) dates to the era of Galileo, Huygens, and Leibnitz. The word "work" has many meanings in any dictionary, of course, but this story starts by introducing the very simplest mechanical definition of work. Simply put, mechanical work increases energy. Its definition (taught by physicists and mechanical engineers) is "work is a movement of something multiplied by the force in the same direction that accompanies that movement". Positive work increases the energy of anything by the precise value of the work done. If a force on an object is known along with a change in location, work can be calculated using this definition. Figure 9.1 shows a simple example that is occasionally used in introductory physics courses that define work and show how it leads to the production of energy. A mass in gravity has weight. Lifting the weight is called work and the lifting to some elevation also increases its potential energy. If we let this elevated mass fall, the energy is converted to speed

Fig. 9.1 How work is converted into potential and kinetic energy

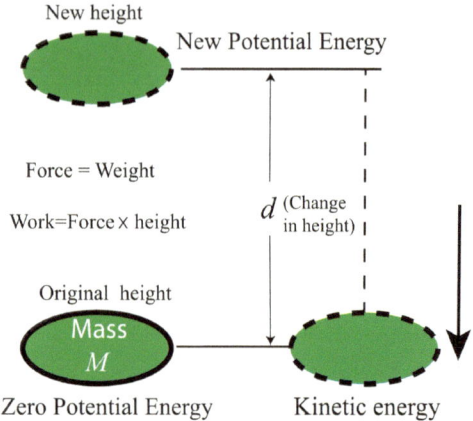

and kinetic energy. Scientifically, it is common in a case like this to say that energy is conserved because the energy has the same value even in different forms. Energy can take many forms and often can be precisely calculated, and such calculations constituted much of my education as an engineer. Consider the vertically displaced mass with gravity present in Fig. 9.1. The rising mass requires work and gains potential energy, and if the mass is released at the top and allowed to fall freely, the potential energy is converted to kinetic energy. We can determine the value of the energy of the mass moving up and down. It leads to the calculation that predicts the speed when it arrives back at the bottom.

Consider lifting a weight with a rope in a field of gravity as in Fig. 9.1. The work is the force of the weight multiplied by the distance lifted. The force of the weight is the mass of the material M times the acceleration of gravity, which has the symbol g. Therefore, using algebra, the force F is calculated to be equal to $M \times g$, and work W to lift it a distance d is easily calculated with the formula $W = M \times g \times d$.

The value of W changes the energy of the object. The important rule for this example is that work equals the increase in potential energy of the weight. Then, if you cut the rope at the top of the lifting process, the weight will fall and at the bottom (if there is no friction) will reach kinetic energy exactly equal to the decrease in potential energy (=work) as it moves from top to bottom. The formula for this potential energy PE is $PE = M \times g \times d$, and the formula for the kinetic energy of a mass M moving with velocity v is $KE = v^2/2$. Equating them allows us to predict the velocity of an object falling from an elevation d. In summary, the work produces an increase in potential energy, and cutting the rope allows acceleration that releases the potential energy and converts it to kinetic energy.

I'd like to emphasize that the definitions of potential and kinetic energy are clearly defined and universal. It applies to a ball, a rock, a basket of oranges, an elephant, an asteroid, or anything else. Consider another case of a ball thrown upward in a field of gravity so that it slows down and reaches a point of zero motion at the elevation d in Fig. 9.1 and then returns down to the origin. Although the final velocity is reversed, the original and final kinetic energy values of the ball are identical. Therefore, there are two values of velocity with the same kinetic energy for the balls. This illustrates an important aspect of energy calculations. Energy calculations alone do not make exact predictions. Knowledge about something else is needed too, like mechanics, chemistry, electromagnetics, biology, or some other structural laws. They provide the skeleton in Fig. 1.1. That is true not only in this example of a mass in a gravity field, but in general. Most energy calculations are far more complicated, and even though the example in Fig. 9.1 is somewhat boring, the simplicity is deceptive. What is important is the fact that the conservation of energy is fundamentally correct. Another thing to note is that this example is idealized, and much of our real world interferes with the perfect calculation. For example, this does not work perfectly if the weight has friction as it moves or if the distance d is so large that the value of g changes. Finally, this example of energy conservation is one of thousands that we might show. There are many forms of energy and of its change and storage. Some things that store energy are batteries, chemicals (including fuels), magnetic fields, nuclear materials, water dams, tanks of hot water, heavy weights lifted high in the air, stars, and spinning objects. They range from being very simple to being very sophisticated.

The example in Fig. 9.1 is taught in classrooms as a part of "Classical Physics", or classical mechanics. Humanity did not always know about this. Discoveries were first made and written down in the 1600s. Experiments could be conducted to verify the results with better and better precision as friction was made smaller and the ability to measure time improved. Energy is fundamentally important, and it is easy to understand why these conversions of energy between different forms were found to be important. When steamboats and trains were developed, for example, the relation between distance traveled, the fuel required, and the frictional forces helped everyone to understand how much fuel was needed and how things worked. Machines were not the only engines, and metals were not the only substances included in the revolutionary understanding of energy. Many other discoveries were added to the mix, such as electric charge, magnetic fields, the conversion of heat to mechanical energy, the conversion of mechanical energy to heat, and the energy trapped within atoms and molecules. We learned to quantify by mathematical expressions how energy is transferred back and forth between mechanical, chemical, electromagnetic, and thermal states. Even light itself transmits energy and all other forms of electromagnetic waves do, too.

This sort of Classical Physics has limitations (ask any physicist). For very small weights and distances, the simple definition of work given above does not agree with experimental results. After much discussion in the nineteenth century, it was found that quantum mechanics required new formulas and mathematical calculations that are unlike those in Fig. 9.1. Another limitation for this sort of Classical Physics occurs for speeds approaching the speed of light or near a great concentration of mass where relativity is needed. After the staggering new developments of relativity and quantum mechanics in the early twentieth century, some people regarded classical physics as occupying a hallowed central spot in physics with not much need for further development.

Ok, so humankind has known about most topics in classical physics for over a hundred years!

So what?

Well, something new that involves classical physics was only understood during the middle of the twentieth century. This "something new" requires the use of the flow of energy, not energy change itself. That story is told here. So here goes…

9.2 Energy Flow and Its Rate

Let's start with a simple example that shows the flow of energy. The definition of work in Fig. 9.1 did not take into account how quickly things occurred, so let's do that now. First, please think of a mass that is forced to move up or down at a steady speed in a field of gravity. Now, the work that is required to move this mass in the presence of this force is done at a constant rate. As time goes on, you can calculate the value of work done over that time interval by taking the total work up to that moment and dividing it by the time interval. This expresses the rate of work that is being done and is called the *power*. This can "work" both ways. Although power at a steady rate is required by a bike rider steadily pedaling up a hill, power also expresses the rate of liberating energy when the bike coasts downhill. Figure 9.2 shows a simple example of that. Dense fluid enters a pipe at its top, flows downward in a field of gravity, and leaves at the bottom. As before, the small-print calculation is an example of how the calculation can be done. The detailed algebra is not needed to understand the concept that power can be calculated.

Fig. 9.2 The power, or the rate of change of energy for fluid in a gravity field descending in a tube at a steady speed while immersed within a lighter fluid

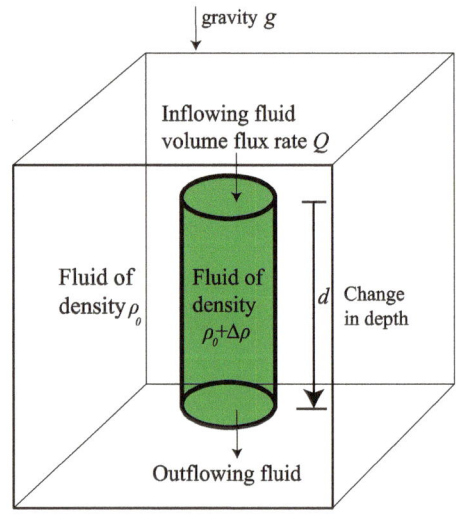

gravity g

Inflowing fluid volume flux rate Q

Fluid of density ρ_0

Fluid of density $\rho_0 + \Delta\rho$

d Change in depth

Outflowing fluid

Rate of change in potential energy $= g \times \Delta\rho \times d \times Q$

Figure 9.2 shows a simple flow that quantifies the power that is released when the fluid that is denser than its surrounding fluid sinks. The tube is immersed in an ambient body of fluid that has a density. Denser fluid, with density $\Delta\rho$ enters the top of a tube and descends a vertical distance where it leaves the bottom of the tube. The rate of fluid descending fluid is the volume of fluid per unit time Q. The rate of potential energy increase is Er that is given by the formula

$$\text{Er} = g \times \Delta\rho \times d \times Q.$$

For readers who are not accustomed to Greek letters, I say, "don't worry, delta and rho are two of the only very few Greek letters used, and each Greek letter will be explained when first used".

Consider as a second example a raindrop starting in a cloud and falling to the ground. Each raindrop rubs air as it falls, so that it gets to a steady falling speed with a balance between the rate of potential energy release and frictional dissipation. Therefore, each drop has the same balance for the rate of change of potential energy as a parcel of water in the pipe shown in Fig. 9.2 so long as the fluid speed within the pipe is constant. For both the raindrop and steady pipe flow, the rate of change in potential energy might be balanced by the rate of heat production by friction. Then, for the raindrop, the kinetic energy at the end of the drop is dissipated by friction whether it hits the ground, a tree, or the ocean. On land, the water percolates through the soil as well as drains into the ocean through streams and rivers. The total energy flow of all the drops starting at some high elevation and falling during a rainstorm before hitting the ground would follow a calculation parallel to that illustrated in Fig. 9.2. Since the raindrops don't accelerate forever, there is some drag, and energy is converted to frictional heat by air drag at a constant rate. Later, there is dissipation during impact and accumulation on the Earth surface or ocean. The water's final kinetic energy is zero. Although the density of the air around a raindrop is much smaller than the density of water surrounding the pipe in Fig. 9.2, that does not alter the algebra.

The example in Fig. 9.2 shows two other features of energy calculations: First, more information is needed than just the energy balance alone to calculate the mechanical problem completely. Examples of important information would be the form of frictional resistance for the pipe flow (or during a raindrop history) or the pathway geometry. Second, other factors can also play a role. In fact, the number of similar examples to Fig. 9.2 and the complications that arise are limitless, because energy converts to different forms.

Despite these limitations, calculations based on energy flow alone can help us. Here is a good example: A 1-year supply of rain probably has almost a constant value with small deviations each year. Taking an estimate for the constant value of rain over our entire planet and using some estimate of the average height of clouds that emit precipitation, the power released by the falling rain on average over the whole year for the entire Earth is a number easily calculated. This simple calculation is not terribly useful as far as I know, because most of the power released by the rainfall is generally lost. That is, it is generally dissipated by the drag of the raindrops through the atmosphere. However, if we build a dam to make electricity in one of the rivers, then some of this power becomes available to us. Knowing the flow of the river upstream of the dam (fed, of course, by rainfall over the upstream watershed of that river) and the drop in elevation as water moves from the upstream lake to below, the dam is extremely useful for estimating how much electricity can be generated.

In general, how does one calculate problems with energy flow? One approach is to start a problem and then calculate what happens as it runs down (like a cloud running out of water). Another approach is to feed the problem with steady forcing and calculate how the energy is converted into various forms. For steady forcing, we balance the input with the energy losses. For example, sunlight shining on Earth produces some evaporation of water leading to the formation of clouds and rain. This cycle involves changing energy at different rates in numerous different forms: radiation; evaporation; condensation; surface energy; potential energy; kinetic energy; and heat. Unfortunately, a calculation of the energy that is lost during an event might not be easy to calculate in detail, because there might be many potential sources of dissipation. Some examples of dissipation are frictional resistance; radiating sound; radiation away by electromagnetic waves; the conduction of heat away from a region, chemical reactions, or turbulence. For the example in Fig. 9.2, one possibility is that the fluid is viscous enough to descend with a buoyancy/friction balance. Another is that it sinks so quickly that it makes turbulent flow within the tube. There can even be an extreme case where very cold air sinking in the tube exceeds the speed of sound. If the fluid is rotating, there might be a vortex centered along the central axis. Turbulence or intense swirling flow with acoustic shock waves might accompany the drag. Since sinking and rising fluid might have the same dynamics except that potential energy rate changes sign, this might be the case for rising air in a tornado, especially if particles are present. Therefore, the actual energy flow is linked not only to the sources supplied at a given rate but by the type of dissipation.

Figure 9.3 shows an example of a simple energy-driven flow. It shows a model of a hot spring or a geyser that propels hot water up to the surface. Let's start with water in a lake that is connected to a buried tube that extends sideways under the lake and finally comes out of the ground nearby at a level slightly higher than the level of the lake surface. The water in the tube is heated at the bottom by geothermal heating so that it gets hot, expands, and acquires a lower density than the lake water. If there is no flow along the tube, the water simply gets hot at the bottom (Fig. 9.3a). If there is flow along the tube (Fig. 9.3b), the water moving sideways becomes warmer and warmer so when it moves toward the surface, it is lighter and buoyantly floats upward. Starting with no flow, the water can stay still forever as in the first case, or it can develop a steady flow as in the second case, and produce a hot spring nearby.

This example has two processes that I mention in the text again and again concerning energy flow. The first process is thermal *conduction*. This is the movement of heat, or thermal energy through material by microscopic movements. Heat is a form of energy stored in any material as vibration. Therefore, hot material (for example, water, rock, magma, or mayonnaise) has molecules that are vibrating more than cold water. When hot material is placed next to colder material, the vibrations spread out from hot to cold, and this is called thermal conduction. In Fig. 9.3, I assume that the hot rock around the buried tube will conduct heat into the water in

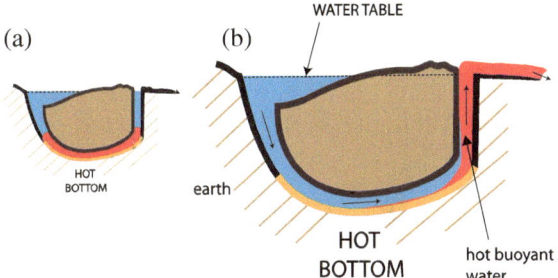

Fig. 9.3 A simple model of flow driven by the flow of heat. A tube extending down through a hot region is filled with water. The high temperature of the region might come, for instance, from volcanic activity. The water can be either **a** motionless or **b** flowing from buoyant driving. In the latter case, cold dense lake water descends in the left-hand part of the tube. It is heated by the hot rocks at the bottom, and the lighter hot water flows to the right and bubbles up as a hot spring

the tube. The rate at which this energy flows is well-known. Roughly speaking, heat flows from hot material to cold material at a rate depending on the thermal conductivity of the material multiplied by the magnitude of the temperature difference. This product is divided by the distance between hot and cold. Thus, a hot place and a cold place 20 miles apart might conduct almost no heat, but the places separated by the same thermal conductivity that are only one inch apart might transport quite a bit.

The second process is called *convection*. Convection means taking a hot material and bringing it to a colder place, where it can warm the colder place. Therefore, if we take the hot rocks around the tube in Fig. 9.3b and pass water close to them, the cold water first heats up by conduction and then it flows to the colder place up higher. The energy in the form of heat is said to be transported by convection. The hot spring has hot water brought to the surface by convection. This is specifically called free convection since the buoyancy force propels the water. We can think of many other examples. Hot stoves have heated air rising around them. The sun makes black pavement hot in the summer, and the rising air heats the atmosphere. Convection is all around us and all around the Earth, too. Engineers must use convection to cool machines quite frequently and if the fluid is pumped, engineers call it forced convection.

The hot water in Fig. 9.3b can even emerge at the ground at a higher level than the source level. An extreme example of this case has the water temperature exceeding the boiling temperature and then a geyser driven by steam jets out. The flow depends on having a sufficiently large temperature difference between the surface and depth so that the buoyancy force from the warmer lighter water balances the frictional drag of the flow as the buoyant fluid flows up and steadily liberates potential energy. Therefore, this example is an upside-down version of the example shown in Fig. 9.2 where dense fluid sinks.

9.3   Quantification of Properties, Units

Before going further, it is important to stress that we humans had to learn about the importance of quantities. The aspects of physics and chemistry involved with energy-flow structures involve values and numbers rather than general concepts alone and we are forced to deal with quantities.

Before using any quantities, let's start with a quick review of measurements and magnitudes. First, we express distance in meters, a bit over 3 feet or 39 in. We express force in Newtons, joules for energy, and kilowatts for the flow of energy. This is called the MKS system of units. It is the international standard, used throughout science now and is principally used in this book to reduce confusion. Many people, groups, and nations use other units. They include units of length (inch, foot, or mile), of force (dyne, kilogram force, pound-force, or poundal), of energy (erg, calorie, British Thermal Unit (BTU), or kilowatt-hour), and of energy flow or power (horsepower, Watts). Occasionally one of those other units will be mentioned for ease in visualization.

The rate of energy change, which can be thought of as the flow of energy from one spot to another with perhaps a change in the form of energy, is called the power. The unit of power is one watt; mechanically, this equals a flow of one joule of energy per second. But in the international MKS system, we use kilowatt. A watt can refer to the flow of form of energy, for example, electric energy, the rate of heat released by combustion, or by mechanical work. Pumping 1/4 of a liter of water up one meter on Earth's surface each second requires one watt. Pumping 250 L of water up one meter each second requires one kilowatt). Virtually, everything described in this book is produced from the flow of energy or the power. If power had not occurred, these things would not exist (me, too).

9.4   Large Numbers

We are almost ready to move on, but first it is necessary to mention the size of things. The things to be described are very different in size from our ordinary life, and we must adopt an expression for very large or small numbers. How do we accomplish this? Lots of times writers describe size by some imaginary trick. For example, the weight of water behind a dam might be expressed as a billion elephants, or the size of a million dollars might be described as around 100 miles with each dollar bill set end to end. We can't communicate effectively using these methods, and it is wisest to use the simplest possible notation to give sizes. We express size by exponents of ten, sometimes called the order of magnitude.

The order of magnitude for a large number like 100 is written as 10^2, for a million we write 10^6, and so forth. This also works for very small numbers. Instead of one divided by a million we write 10^{-6}, and so forth. The order of magnitude notation makes the relative sizes of different things obvious. For example, 10^{12} is much bigger than 10^6, and 10^{-12} is much smaller than 10^{-6}.

To be a bit more precise than a factor of ten, these numbers can be multiplied by a value in front of the ten. Here is an example. The size of energy generated in 2021 for the United States is described by the U. S. Energy Information Administration website to be roughly "97,331,601,000,000,000 Btu, or about 97 quadrillion Btu" (Actually, what they mean, but don't say, is that this is the annual consumption of energy produced by the combustion of fuels for the entire year). Using the order of magnitude, the value is 9.7331601×10^{16} BTU per year. But I want to be even simpler. So far, this is not using the international units of joules for energy and seconds for time, but this is easily converted. First, we convert BTU to joules and get a rounded off value of 1.02×10^{20} J per year, and then we convert the time units from years to seconds. Since a year is $365.25 \times 24 \times 3600$ s, we find the value is 3.23×10^{12} W or 3.23×10^9 kW. This amount is about 16% of the worldwide human consumption that is approximately 20×10^9 kW. In previous chapters, we have compared 3.23×10^{12} W with energy-flow values of the giant engines.

9.5 Dimensionless Numbers

Generally, a dimensionless number expresses balances. The Rayleigh number can be thought of as a balance between the force of buoyancy and viscous drag force. In other cases, a dimensionless number is a balance between two-time scales or two length scales. What does that mean? Simply put, a time scale is the time that it takes for something to happen. If you fill a pail of water with a hose, the filling time scale is the volume of the pail divided by the volume flow rate coming out of the hose. For example, filling a ten-gallon pail with a flow of one gallon per minute takes a time scale of ten minutes. A length scale is a distance for something to change. If you throw a pebble into a still pond, the radius of the wave pattern around the source steadily increases as the waves spread out in an expanding circle. A dimensionless number is the radius at some instant after the pebble arrives divided by the propagation speed of the wave pattern times time. One of the most important implications of a dimensionless number arises if a length is included within a dimensionless number, because in that case the size of the application can be "scaled" to different sizes. This possibility is common in fluid mechanics.

We use the word "scaled" to mean re-scaled from one problem to another if they have the same dimensionless number. Therefore, a meteorite hitting the ocean might make a pattern identical to the pebble, but it can be scaled up to the ocean size or down to a much smaller size (depending on the size of the meteorite). The same dimensionless number applies no matter how large or small. Likewise, if time is included, properties can be scaled to different time intervals. Therefore, in cellular convection, the convection layer has a depth (the letter d in the calculation given above), so it can represent various applications with different sizes. Some examples are the layer of liquid in a kettle of soup, the depth of a planetary atmosphere, mantle convection discussed in Chap. 3, or a very tiny layer in a microscopic.

Dimensionless numbers in fluid dynamics and solid mechanics are widespread and allow engineers to test aircraft designs in wind tunnels, to test scaled-down ship designs in towing tanks, and to test bridge and building designs in architecture. Scaling with the use of dimensionless numbers is fundamental to numerical models. This is exactly the reason why computer programs work for weather forecasting, climate modeling, and many other physical processes. To help us to understand the Earth, we apply the principle of scaling by using dimensionless numbers to analyze the mechanics and energetics to widespread applications with different sizes in the natural world.

A simple example shows how scaling is used in conjunction with dimensionless numbers. So far in this narrative, the principle that governs speed is explained using conservation of energy and the example under Fig. 9.2 produces a calculated velocity equal to $2\sqrt{g \times \alpha \times \Delta T \times d}$. This formula has the dimensions of distance divided by time, (velocity). A situation with these parameters can have fixed values when the problem is specified and the formula can define a velocity scale. If you take any physical velocity (perhaps calculated from numerical calculations or measured in experiments) and divide that speed by the velocity scale, then that combination is dimensionless (a velocity divided by our formula with the dimensions of velocity). The dimensionless velocity is said to be "scaled".

Much effort has been spent to find an experiment with cellular convection with the measured speed as large as the velocity scale in the above formula! Speeds are generally not measured, but the bulk heat flow is, and this scaling is consistent with a heat flow that is proportional to $Ra^{1/2}$. The thought for many years was that this scaling law might exist for extremely large values of Ra. Although the largest values possible have not found this heat flow relation, five years ago it was found for heat flow driven convection (Lepot et al. 2018) did obey that scaling even though values of Ra are only moderate. The explanation was that heat flow driven convection is not impeded by flows near the top and bottom boundaries. I am personally very excited by this result. The speed measurements are not reported, and I hope that they will be someday.

For cellular convection, both vertical and horizontal distances can be made dimensionless by dividing by fluid depth d. Time can also made dimensionless by multiplying the time by thermal diffusivity κ (this has the dimensions of length squared divided by time (m^2/s)) and then dividing by the square of the depth of the layer d (with dimensions of length (m)). Dimensionless length, velocity, and time are used in the figures in the rest of this chapter and throughout the book.

Dimensionless numbers are notoriously challenging for students to understand at first. It is not their fault, because dimensionless numbers are relatively rare, and in many cases, there is not one unique cluster of dimensionless numbers. Things are even more confusing because most situations can be expressed by several different dimensionless numbers. The issue is further complicated because dimensionless numbers are usually named after those people who first used them, so that naming them depends on who records history. Sometimes one group of engineers and scientists uses one name, and another group uses another name. Names also vary in different countries. In other cases, one dimensionless number is used for some problems and its inverse is used for another. These confusing aspects are kept to a minimum in this book.

The basis of dimensionless numbers lies in the fact that physical units are special. For example, a length is simply that and a unit of time is simply that, too. Matter has mass, electricity has electric charge, and magnetism has a strength, too. As explained in Chap. 1, each of these has a unit. Each physical system has combinations of these, for example, water has a density that is mass per unit volume. A moving ball has both momentum (mass times velocity) and kinetic energy (mass time velocity squared). Flowing water has density, viscosity, and speed. Those three properties are often combined and expressed as a dimensionless Reynolds number (velocity times a length times a density divided by viscosity). A theorem exists, which we don't review saying that the number of dimensionless numbers equals the number of parameters divided by the dimensions. Thus, the parameters of the flow around a ball moving through the air (for example) are the radius of the ball, the density of the air, the viscosity of the air, and the speed, a total of 4. The dimensions are length, time, and force, a total of 3, so one combination exists, and the ultimate combination is one number, the Reynolds number. There is no other way to make a dimensionless number except of course to use its inverse.

In this book, the Rayleigh number is used frequently. It has more parameters than Reynolds number. The parameters are depth; temperature difference; the force of gravity, thermal conduction; viscosity; density; specific heat; and coefficient of expansion. This makes eight parameters, and the dimensions are length, time, temperature, heat, and force. Subtracting, $8 - 5 = 3$, so three dimensionless numbers exist. We can simplify a bit and reduce the number of dimensions. When we combine thermal conductivity, specific heat, and density, we make a single variable called thermal diffusivity (eliminating heat). We can also divide the value of viscosity by density to make a variable called viscous diffusivity (eliminating mass or force). Now there are six parameters (depth, a temperature difference, the force of gravity, thermal diffusivity, viscous diffusivity, and coefficient of expansion) and three dimensions (temperature, length, and time) Therefore, it is still a fact that three dimensionless numbers exist. The three common dimensionless numbers are the Rayleigh number, the Prandtl number (ratio of viscous and thermal diffusivities), and the relative change in density (coefficient of expansion times temperature difference). Dimensionless numbers with different names can be found by combining these. One example is the Grashof number. It is the Rayleigh divided by the Prandtl number. It is used for problems where an external flow is imposed.

To summarize this section, I hope that these instructions are useful to people who are less specialized than me. The language and implicit understanding possessed by a trained specialist like me challenges my communication to the broader world, and I hope this set of tutorials helps.

Reference

Lepot S, Aumaître S, Gallet B (2018) Radiative heating achieves the ultimate regime of thermal convection. Proc Natl Acad Sci 115(36):8937–8941

Chapter 10
Summary

The book starts by describing the flow of energy. Then the great discovery of seafloor spreading is narrated. It that led to the general realization that the interior of the Earth was moving. The motion took the form of great plates that spread apart and plunged back into the Earth. The structure of the motion resembles to a limited extent a form of fluid motion called cellular convection that had been studied by physicists and engineers. The suggestion is advanced that these giant forms of flow are driven and maintained by the flow of energy.

How did this flow pattern originate? Chapter 2 describes how we think it happened. First we have to understand a very simple energy-flow structure. It is a self-organizing flow pattern that was extensively studied from the mid 1950s onward called convection cells. This pattern of flow emerges spontaneously. Scientists found how to describe these cells and their behavior using readily accepted calculations and how to verify the behavior with controlled, reproducible experiments. They followed the strict rules of natural philosophy—the understanding and calculations must favorably compare with experiments (or at least, physical measurement of the cells in some natural setting), and the agreement between theory and measurements is the ultimate criterion of their worth. The circulating cell structure consists of simple upward and downward thermal plumes and closed circulation flows that are driven by heat flow, and so they qualify as energy-flow structures. The chapter also tells how these simple experiments can be related to flows that are orders of magnitude larger or smaller.

Chapter 3 quantifies the giant engine of mantle convection. The plates are driven within the Earth by an energy flow that is very similar to the laboratory and theoretical convection cells in Chap. 2. The overall structure of the circulation has dramatic similarities and distinct differences to the convection cells described in Chap. 2 that exist in a fluid dynamics laboratory. Despite the differences, both convection cells and mantle convection are energy-flow structures driven by the flow of thermal energy (heat) upwards toward Earth's surface.

Chapter 4 describes magma flows. These energy-flow structures originate from deep in the Earth starting with a solid melting to a liquid, then with liquid flowing

J. A. Whitehead, *Energy Flow and Earth*, SpringerBriefs in Earth System Sciences, https://doi.org/10.1007/978-3-031-62694-4_10

up and solidifying at or near the cold top of the mantle. Three distinct geological regions with magma flow are described. Ocean ridges are the region with the greatest total magma flow. Island arcs are the second region, hotspots are the third. The total magnitude of heat flow decreases in sequence for these three, and all of them have a smaller total heat flow delivered to the top than the vertical circulation of the plates in Chap. 3. All the magma flows are doubly important because they have been the source of large chemical differentiation near the surface of planet Earth. Magma flow has provided most of the material that makes the continents. It also has brought up the carbon dioxide, the steam (water), the nitrogen for our air, and virtually all the other elements that have migrated from the inside of the Earth to the surface.

Chapter 5 explains how cellular convection and plate motions have all led to the miraculous result that almost all the water on our planet occupies the ocean basins. Our understanding of the gradual formation and the evolution of continents and ocean basins over time is still in an early stage of development, but the story of what is known so far is fascinating. To form the continents, both plate motions and magmas modify and rework the material that we call continental rocks that have lower density than the mantle. Continents float on the mantle just like ice floating on water. This chapter has calculations of a simple model of continent and ocean basin formation that show how to understand the form of the continents, and how both mountain building, and erosion are important.

Chapter 6 tells of another energy-flow structure on Earth that became much more clearly understood in my lifetime. Magnetic fields are generated in planets and stars. The field develops out of nothing. All that is needed is a flowing fluid that can conduct electricity but it has to flow rapidly enough to get the field going. Therefore, it has a criterion for generation. The flow of an electrically conducting fluid that converts the kinetic energy of the flow to energy within a magnetic field was developed by numerous friends and colleagues as I watched in wonder and awe.

Chapter 7 looks at the winds and oceans. First the atmosphere. The wind belts of Earth are fundamental to the climate and weather that concerns us all. There are two pairs of great belts of winds: the tropical trade winds called easterlies (from the east) and the mid-latitude westerlies. The mechanical explanation of the trade winds as a form of convection modified by the rotation of Earth is quite simple, but the westerlies require instability and transitions to form in their present locations halfway between equators and poles. Therefore, the famous meandering jet stream can be considered an energy-flow structure. In addition, clouds are energy-flow structures that form and vanish again. They involve a phase change of water vapor to liquid. They produce rain and they have heat flow as a key component. Atmospheric winds drive the ocean currents near the surface. Many ocean currents are simply driven by the wind and are not strictly speaking energy-flow structures but within them are eddies and gyres produced by instability that are energy-flow structures. Oceanic flows contribute up to 50% of the total heat flow from tropical to polar regions. Therefore, they are a key component in the energy flows of our climate that supports all of us. Some aspects of the wind-driven ocean circulation were discovered during my lifetime, but the understanding of eddies and the buoyancy-driven flows in the ocean has a number of factors that still baffle me.

Chapter 8 comments on the social use of energy. All energy-flow structures have much in common with each other. They rely on the fact that there is dissipation that balances the energy flow that produces the structure. Everything in nature: animals, plants, microorganisms, and perhaps viruses are all energy-flow structures. Every living thing itself has the same energetic features as those in Fig. 1.3. There can be either life (motion) or death (no motion). The same can be said of commerce and civilizations. Therefore a case study of energy use by society is reviewed.

Chapter 9 is a tutorial that reviews the discovery of energy and its conservation along with the definitions of kinetic and potential energy, and then gives a simple example of the rate of production of energy for buoyantly driven fluid flow. The quantification of energy flow is reviewed along with an explanation of orders of magnitude. An example is given of a fluid flow driven by energy flow, notably a water flow driven by the heat flow into a tube below a lake leading to a hot spring. Finally dimensionless numbers, the units used in this book, and the order of magnitudes are briefly explained.

We have stepped back mentally and looked at the large-scale role of energy flow on Earth. We recognize the importance of energy flow in constructing and maintaining many large structures. These include mantle convection, volcanic flows, the formation of continents and ocean basins, the magnetic field, our weather, Earth's climate, and ocean circulation. The underlying structural skeletons are explained by the dynamics that mankind has developed to understand these phenomena, but energy flow is the fuel that keeps it all going. I hope that people working in many branches of arts and sciences can visualize the role of energy more fully because of this book. The final objective is to use science, math, and ideas to understand our home, Earth.